SOIL BIOCHEMISTRY

SOIL BIOCHEMISTRY

By

Vijay Mittal

2012

SBS Publishers & Distributors Pvt. Ltd.

New Delhi

ISBN 13 : 9789380090528

First Published in 2012

Published by:

SBS PUBLISHERS & DISTRIBUTORS PVT. LTD.
2/9, Ground Floor, Ansari Road, Darya Ganj,
New Delhi - 110002,
INDIA
Tel: 0091.11.23289119 / 41563911 / 32945311
Email: mail@sbspublishers.com
www.sbspublishers.com

Printed in India by Chaman Enterprises, New Delhi.

Preface

Soils are a complex mixture of air, water, organic and inorganic solids. Soil biochemistry is a branch of soil science that is concerned with the biochemical reactions involving these phases. Traditional soil biochemistry has focused primarily on chemical reactions in the soil that affect plant growth. For example, a great deal of research has been conducted on N-P-K chemistry in soils in order to increase crop yields and fertilizer efficiency. However, during the past several decades there have been growing concerns about organic and inorganic contamination of important resources and potential ecological and human health risks. Consequently, the emphasis in soil biochemistry has shifted from strictly agricultural to environmental. Knowledge of environmental soil biochemistry is paramount to predicting the fate, mobility and potential toxicity of contaminants in the environment. The vast majority of environmental contaminants are initially released to the soil. Once a chemical is exposed to the soil environment a myriad of chemical reactions can occur that may increase/decrease contaminants toxicity. These reactions include adsorption/desorption, precipitation, polymerization, dissolution, complication, and oxidation/reduction. These reactions are often disregarded by scientists and engineers involved with environmental remediation. Understanding these processes will enable us to better predict the fate and toxicity of contaminants and will ultimately provide us with the knowledge to develop sound and cost-effective remediation strategies.

Author

Content

Chapter 1

Introduction

SOIL DESCRIPTION AND CLASSIFICATION

It is necessary to adopt a formal system of soil description and classification in order to describe the various materials found in ground investigation. Such a system must be comprehensive (covering all but the rarest of deposits), meaningful in an engineering context *(so that engineers will be able to understand and interpret)* and yet relatively concise. It is important to distinguish between description and classification: Description of soil is a statement describing the physical nature and state of the soil.

It can be a description of a sample, or a soil *in situ*. It is arrived at using visual examination, simple tests, observation of site conditions, geological history, etc. Soil classification is the separation of soil into classes or groups each having similar characteristics and potentially similar behaviour. A classification for engineering purposes should be based mainly on mechanical properties, *e.g.* permeability, stiffness, strength. The class to which a soil belongs can be used in its description.

BASIC CHARACTERISTICS OF SOILS

Soils consist of grains (mineral grains, rock fragments, etc.) with water and air in the voids between grains. The water and air contents are readily changed by changes in conditions and location: soils can be perfectly dry (have no water content) or be fully saturated (have no air content) or be partly saturated (with both air and water present). Although the size and shape of the solid (granular) content rarely changes at a given point, they can vary considerably from point to point.

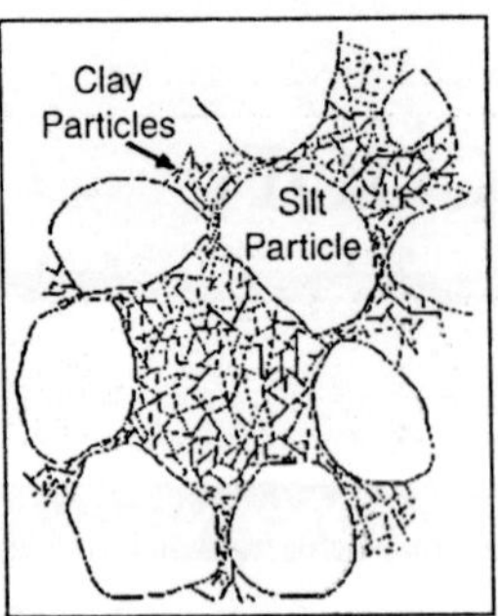

Fig. 1.1

First of all, consider soil as a engineering material-it is not a coherent solid material like steel and concrete, but is a particulate material. It is important to understand the significance of particle size, shape and composition, and of a soil's internal structure or fabric.

Soil as an Engineering Material

The term "soil" means different things to different people: To a geologist it represents the products of past surface processes. To a pedologist it represents currently occurring physical and chemical processes. To an engineer it is a material that can be: built on: foundations to buildings, bridges. built in: tunnels, culverts, basements. built with: roads, runways, embankments, dams. supported: retaining walls, quays. Soils may be described in different ways by different people for their different purposes. Engineers' descriptions give engineering terms that will convey some sense of a soil's current state and probable susceptibility to future changes (*e.g.* in loading, drainage, structure, surface level). Engineers are primarily interested in a soil's mechanical properties: strength, stiffness, permeability. These depend primarily on the nature of the soil grains, the current stress, the water content and unit weight.

Size Range of Grains

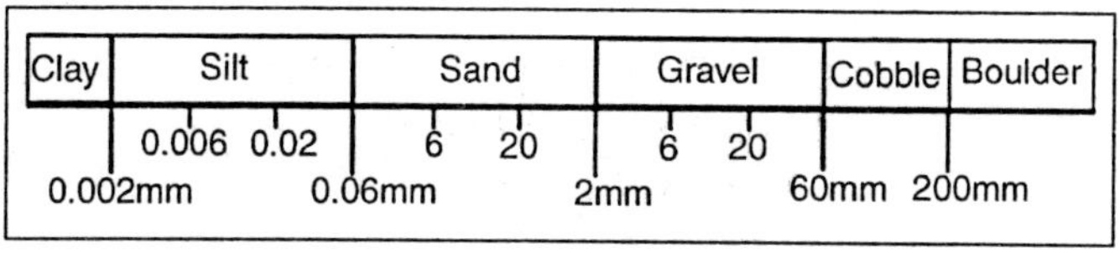

Clay	Silt	Sand	Gravel	Cobble	Boulder
	0.006 0.02	6 20	6 20		

0.002mm 0.06mm 2mm 60mm 200mm

Fig. 1.2

The range of particle sizes encountered in soil is very large: from boulders with a controlling dimension of over 200mm down to clay particles less than 0.002mm (2μm). Some clays contain particles less than 1 mm in size which behave as colloids, *i.e.* do not settle in water due solely to gravity. In the British Soil Classification System, soils are classified into named Basic Soil Type groups according to size, and the groups further divided into coarse, medium and fine sub-groups:

Very coarse soils	Boulders		> 200 mm
	Cobbles		60 - 200 mm
Coarse soils	G Gravel	Coarse	20 - 60 mm
		Medium	6 - 20 mm
		Fine	2 - 6 mm
	S Sand	Coarse	0.6 - 2.0 mm
		Medium	0.2 - 0.6 mm
		Fine	0.06 - 0.2 mm
Fine soils	M Silt	Coarse	0.02 - 0.06 mm
		Medium	0.006 - 0.02 mm
		Fine	0.002 - 0.006 mm
	C Clay		< 0.002 mm

Aids to Size Identification

Soils possess a number of physical characteristics which can be used as aids to size identification in the field. A handful of soil rubbed through the fingers can yield the following:

Sand (and coarser) particles are visible to the naked eye. SILT particles become dusty when dry and are easily brushed off hands and boots. Clay particles are greasy and sticky when wet and hard when dry, and have to be scraped or washed off hands and boots.

SHAPE OF GRAINS

- Shape characteristics of Sands grains
- Shape characteristics of Clay grains
- Specific surface

The majority of soils may be regarded as either SANDS or Clay:

Sands include gravelly sands and gravel-sands. Sand grains are generally broken rock particles that have been formed by physical weathering, or they are the resistant components of rocks broken down by chemical weathering. Sand grains generally have a rotund shape.

Clays include silty clays and clay-silts; there are few pure silts (*e.g.* areas formed by windblown Löess). Clay grains are usually the product of chemical weathering or rocks and soils. Clay particles have a flaky shape.

THERE ARE MAJOR DIFFERENCES INCHARACTERISTICS OF SAND GRAINS

Sand and larger-sized grains are rotund. Coarse soil grains (silt-sized, sand-sized and larger) have different shape characteristics and surface roughness depending on the amount of wear during transportation (by water, wind or ice), or after crushing in manufactured aggregates. They have a relatively low specific surface (surface area).

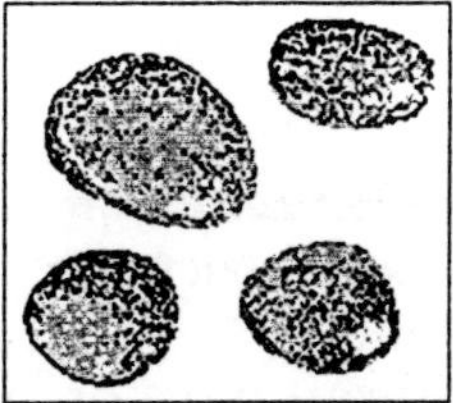

Fig. 1.3 Rounded: Water- or Air-Worn; Transported Sediments

Fig. 1.4 Irregular: Irregular Shape With Round Edges; Glacial Sediments (Sometimes Sub-Divided into 'Sub-Rounded' and 'Sub-angular')

Fig. 1.5 Angular: Flat Faces and Sharp Edges; Residual Soils, Grits

Fig. 1.6 Flaky: Thickness Small Compared to Length/Breadth; Clays

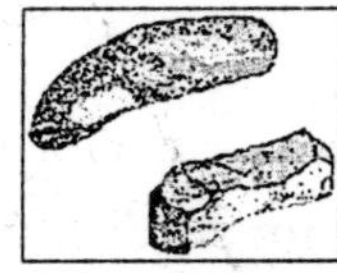

Fig. 1.7 Elongated: Length Larger Than Breadth/Thickness; Scree, Broken Flagstone

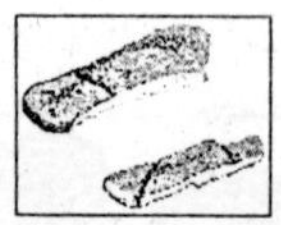

Fig. 1.8 Flaky and Elongated: Length>Breadth>Thickness; Broken Schists and Slates

SHAPE CHARACTERISTICS OF CLAY GRAINS

CLAY particles are flaky. Their thickness is very small relative to their length and breadth, in some cases as thin as 1/100th of the length. They therefore have high to very high specific surface values. These surfaces carry a small negative electrical charge, that will attract the positive end of water molecules. This charge depends on the soil mineral and may be affected by an electrolite in the pore water. This causes some additional forces between the soil grains which are proportional to the specific surface. Thus a lot of water may be held asadsorbed water within a clay mass.

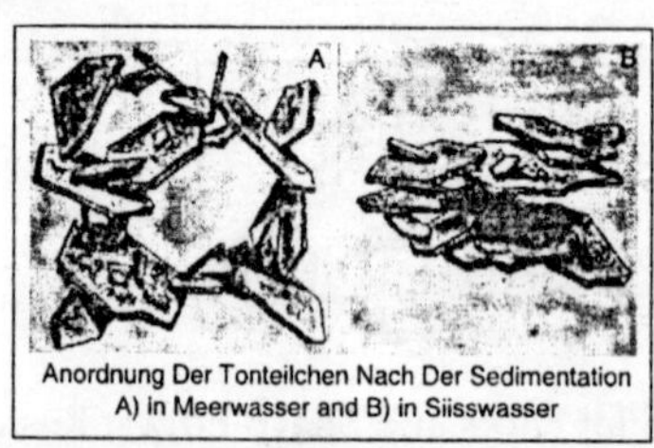

Anordnung Der Tonteilchen Nach Der Sedimentation A) in Meerwasser and B) in Siisswasser

Fig. 1.9

SPECIFIC SURFACE

Examples

Specific surface is the ratio of surface area per unit wight. Surface forces are proportional to surface area (*i.e.* to d^2).

Hence, specific surface is a measure of the relative contributions of surface forces and self-weight forces. The specific surface of a 1mm cube of quartz (ρ = 2.65gm/cm^3) is 0.00023 m^2/N Sand grains (size 2.0–0.06mm) are close to cubes or spheres in shape, and have specific surfaces near the minimum value. Clay particles are flaky and have much greater specific surface values.

Examples of Specific Surface

The more elongated or flaky a particle is the greater will be its specific surface.

Cubes

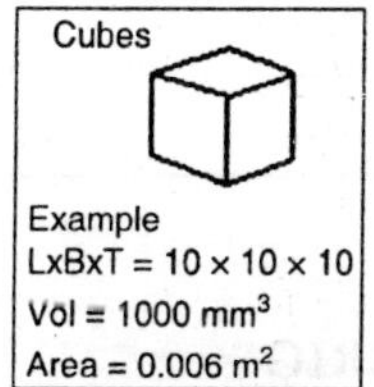

Fig. 1.10

Rods

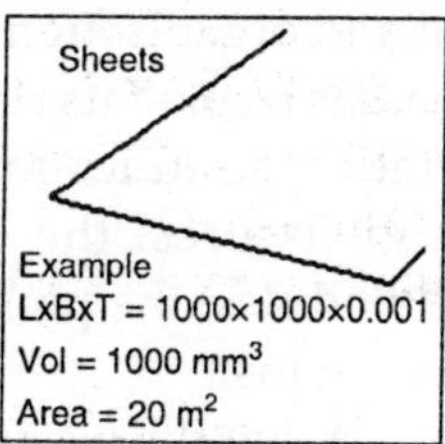

Fig. 1.11

Sheets

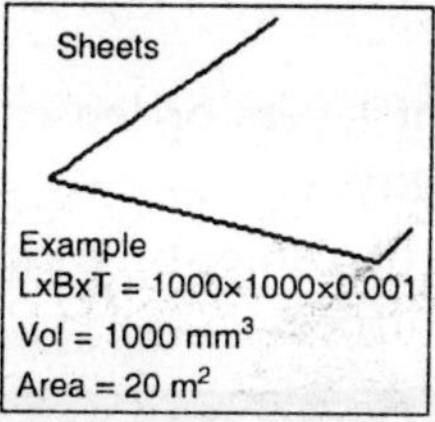

Fig. 1.12

Table. Examples of Mineral Grain Specific Surfaces

Mineral/Soil	Grainwidth d (μm)	Thickness	Specific Surface m^2/N
Quartz grain	100	d	0.0023
Quartz sand	2.0–0.06	≈d	0.0001–0.004
Kaolinite	2.0–0.3	≈0.2d	2
Illite	2.0–0.2	≈0.1d	8
Montmorillonite	1.0–0.01	≈0.01d	80

STRUCTURE OR FABRIC

Natural soils are rarely the same from one point in the ground to another. The content and nature of grains varies, but more importantly, so does the arrangement of these. The arrangement and organisation of particles and other features within a soil mass is termed its structure or fabric. This includes bedding orientation, stratification, layer thickness, the occurrence of joints and fissures, the occurrence of voids, artefacts, tree roots and nodules, the presence of cementing or bonding agents between grains.

Structural features can have a major influence on in situ properties:

- Vertical and horizontal permeabilities will be different in alternating layers of fine and coarse soils.
- The presence of fissures affects some aspects of strength.
- The presence of layers or lenses of different stiffness can affect stability.
- The presence of cementing or bonding influences strength and stiffness.

ORIGINS, FORMATION AND MINERALOGY

Soils are the results of geological events (except for the very small amount produced by man).

The nature and structure of a given soil depends on the geological processes that formed it:

- *Breakdown of parent rock:* Weathering, decomposition, erosion.
- *Transportation to site of final deposition:* Gravity, flowing water, ice, wind.
- *Environment of final deposition:* Flood plain, river terrace, glacial moraine, lacustrine or marine.
- *Subsequent conditions of loading and drainage:* Little or no surcharge, heavy surcharge due to ice or overlying deposits, change from saline to freshwater, leaching, contamination.

ORIGINS OF SOILS FROM ROCKS

All soils originate, directly or indirectly, from solid rocks in the Earth's crust:

- *Igneous Rocks*: Crystalline bodies of cooled magma *e.g.* granite, basalt, dolerite, gabbro, syenite, porphyry
- *Sedimentary Rocks*: Layers of consolidated andcemented sediments, mostly formed in bodies of water (seas, lakes, etc.) *e.g.* limestone, sandstones, mudstone, shale, conglomerte
- *Metamorphic Rocks*: Formed by the alteration of existing rocks due to heat from igneous intrusions (*e.g.* marble, quartzite, hornfels) or pressure due to crustal movement (*e.g.* slate, schist, gneiss).

WEATHERING OF ROCKS

Physical Weathering

Physical or mechanical processes taking place on the Earth's surface, including the actions of water, frost, temperature changes, wind and ice; cause disintegration and wearing.

The products are mainly coarse soils (silts, sands and gravels). Physical weathering produces Very Coarse soils and Gravels consisting of broken rock particles, but Sands and Silts will be mainly consists of mineral grains.

Chemical Weathering

Chemical weathering occurs in wet and warm conditions and consists of degradation by decomposition and/or alteration. The results of chemical weathering are generally fine soils with separate mineral grains, such as Clays and Clay-Silts.

The type of clay mineral depends on the parent rock and on local drainage. Some minerals, such as quartz, are resistant to the chemical weathering and remain unchanged.

Quartz

A resistant and enduring mineral found in many rocks (*e.g.* granite, sandstone). It is the principal constituent of sands and silts, and the most abundant soil mineral. It occurs as equidimensional hard grains.

Haematite

A red iron (ferric) oxide: resistant to change, results from extreme weathering. It is responsible for the widespread red or pink colouration in rocks and soils. It can form a cement in rocks, or a duricrust in soils in arid climates.

Micas

Flaky minerals present in many igneous rocks. Some are resistant, *e.g.* muscovite; some are broken down, *e.g.* biotite.

Clay Minerals

These result mainly from the breakdown of feldspar minerals. They are very flaky and therefore have very large surface areas. They are major constituents of clay soils, although clay *soil* also contains silt sized particles.

CLAY MINERALS

Clay minerals are produced mainly from the chemical weathering and decomposition of feldspars, such as orthoclase and plagioclase, and some micas.

They are small in size and very flaky in shape. The key to some of the properties of clay soils, *e.g.* plasticity, compressibility, swelling/shrinkage potential, lies in the structure of clay minerals.

Fig. 1.13

There are three main groups of clay minerals:

1. *Kaolinites:* (Include kaolinite, dickite and nacrite) formed by the decomposition of orthoclase feldspar (*e.g.* in granite); kaolin is the principal constituent in china clay and ballclay.
2. *Illites:* (Include illite and glauconite) are the commonest clay minerals; formed by the decomposition of some micas and feldspars; predominant in marine clays and shales.
3. *Montmorillonites:* (Also called smectites or fullers' earth minerals) (include calcium and sodium momtmorillonites, bentonite and vermiculite) formed by the alteration of basic igneous rocks containing silicates rich in Ca and Mg; weak linkage by cations (*e.g.* Na^+, Ca^{++}) results in high swelling/shrinking potential

TRANSPORTATION AND DEPOSITION

The effects of weathering and transportation largely determine the basic nature of the soil (*i.e.* the size, shape, composition and distribution of the grains).

The environment into which deposition takes place, and subsequent geological events that take place there, largely determine the state of the soil, (*i.e.* density, moisture content) and the structure or fabric of the soil (*i.e.* bedding, stratification, occurrence of joints or fissures, tree roots, voids, etc.)

Transportation

Due to combinations of gravity, flowing water or air, and

moving ice. In water or air: grains become sub-rounded or rounded, grain sizes are sorted, producing poorly-graded deposits.

In moving ice: grinding and crushing occur, size distribution becomes wider, deposits are well-graded, ranging from rock flour to boulders.

Deposition

In flowing water, larger particles are deposited as velocity drops, *e.g.* gravels in river terraces, sands in floodplains and estuaries, silts and clays in lakes and seas. In still water: horizontal layers of successive sediments are formed, which may change with time, even seasonally or daily.

- *Deltaic and shelf deposits:* Often vary both horizontally and vertically.
- From glaciers, deposition varies from well-graded basal tills and boulder clays to poorly-graded deposits in moraines and outwash fans.
- *In arid conditions:* Scree material is usually poorly-graded and lies on slopes.
- Wind-blown Loess is generally uniformly-graded and false-bedded.

LOADING AND DRAINAGE HISTORY

The current state (*i.e.* density and consistency) of a soil will have been profoundly influenced by the history of loading and unloading since it was deposited.

Changes in drainage conditions may also have occurred which may have brought about changes in water content.

Loading/Unloading History

Initial Loading

During deposition the load applied to a layer of soil increases as more layers are deposited over it; thus, it is compressed and water is squeezed out; as deposition continues, the soil becomes stiffer and stronger.

Unloading

The principal natural mechanism of unloading is erosion of overlying layers. Unloading can also occur as overlying ice-sheets and glaciers retreat, or due to large excavations made by man. Soil expands when it is unloaded, but not as much as it was initially compressed; thus it stays compressed-and is said to be overconsolidated. The degree of overconsolidation depends on the history of loading and unloading.

Drainage History

Chemical Changes

Some soils initially deposited loosely in saline water and then inundated with fresh water develop weak collapsing structure. In arid climates with intermittent rainy periods, cycles of wetting and drying can bring minerals to the surface to form a cemented soil.

Climate Changes

Some clays (*e.g.* montmorillonite clays) are prone to large volume changes due to wetting and drying; thus, seasonal changes in surface level occur, often causing foundation damage, especially after exceptionally dry summers. Trees extract water from soil in the process of evapotranspiration; The soil near to trees can therefore either shrink as trees grow larger, or expand following the removal of large trees.

GRADING AND COMPOSITION

The recommended standard for soil classification is the British Soil Classification System, and this is detailed in BS 5930 Site Investigation.

COARSE SOILS

- Particle size tests
- Typical grading curves
- Grading characteristics
- Sieve analysis example

Table. Coarse Soils Are Classified Principally on The Basis of Particle Size and Grading.

Very coarse soils	Boulders		> 200 mm
	Cobbles		60 - 200 mm
Coarse soils	**G** Gravel	Coarse	20 - 60 mm
		Medium	6 - 20 mm
		Fine	2 - 6 mm
	S Sand	Coarse	0.6 - 2.0 mm
		Medium	0.2 - 0.6 mm
		Fine	0.06 - 0.2 mm

Particle Size Tests

The aim is to measure the distribution of particle sizes in the sample. When a wide range of sizes is present, the sample will be sub-divided, and separate tests carried out on each sub-sample. Full details of tests are given in BS 1377: "Methods of test for soil for civil engineering purposes".

Fig. 1.14

Particle-Size Tests

- *Wet sieving:* To separate fine grains from coarse grains is carried out by washing the soil specimen on a 60mm sieve mesh.

- *Dry sieving:* Analyses can only be carried out on particles > 60 mm. Samples (with fines removed) are dried and shaken through a nest of sieves of descending size.
- *Sedimentation:* Is used only for fine soils. Soil particles are allowed to settle from a suspension. The decreasing density of the suspension is measured at time intervals. Sizes are determined from the settling velocity and times recorded. Percentages between sizes are determined from density differences.

Particle-Size Analysis

The cumulative percentage quantities finer than certain sizes (*e.g* passing a given size sieve mesh) are determined by weighing. Points are then plotted of % finer (passing) against log size. A smooth S-shaped curve drawn through these points is called a grading curve. The position and shape of the grading curve determines the soil class. Geometrical grading characteristics can be determined also from the grading curve.

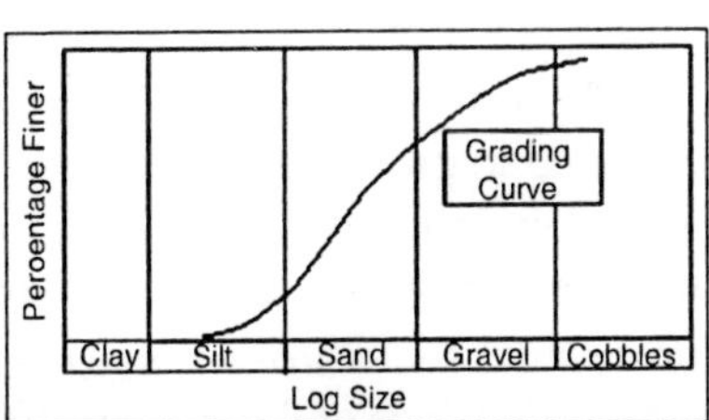

Fig. 1.15

Typical Grading Curves

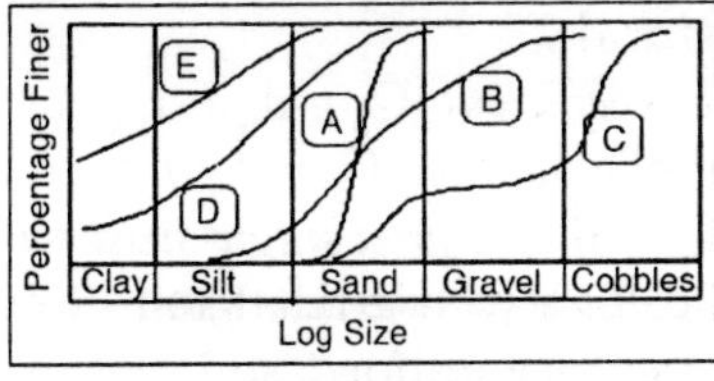

Fig. 1.16

Both the position and the shape of the grading curve for a soil can aid its identity and description.

Some typical grading curves are shown in the figure 1.16:

- A–a poorly-graded medium Sand (probably estuarine or flood-plain alluvium)
- B–a well-graded Gravel-Sand (*i.e.* equal amounts of gravel and sand)
- C–a gap-graded Cobbles-Sand
- D–a sandy Silt (perhaps a deltaic or estuarine silt)
- E–a typical silty Clay

Grading Characteristics

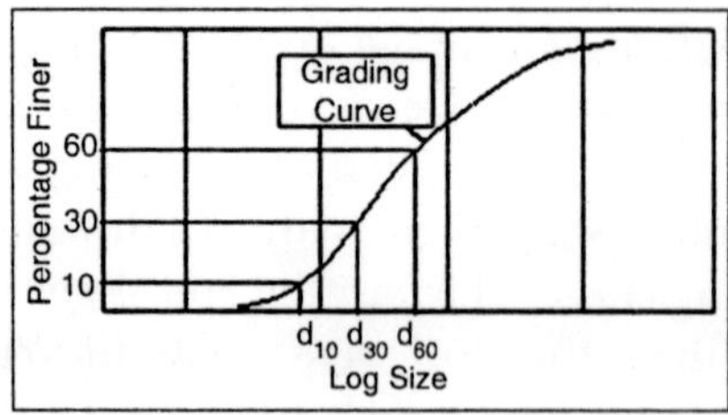

Fig. 1.17

A grading curve is a useful aid to soil description. Grading curves are often included in ground investigation reports. Results of grading tests can be tabulated using geometric properties of the grading curve. These properties are called grading characteristics

First of all, three points are located on the grading curve:

1. d_{10} = The maximum size of the smallest 10% of the sample
2. d_{30} = The maximum size of the smallest 30% of the sample
3. d_{60} = The maximum size of the smallest 60% of the sample

From these the grading characteristics are calculated:

- *Effective size:* d_{10}
- *Uniformity coefficient:* $C_u = d_{60} / d_{10}$
- *Coefficient of gradation:* $C_k = d_{30}^2 / d_{60}\, d_{10}$

Both C_u and C_k will be 1 for a single-sized soil:

- $C_u > 5$ indicates a well-graded soil
- $C_u < 3$ indicates a uniform soil
- C_k between 0.5 and 2.0 indicates a well-graded soil

- $C_k < 0.1$ indicates a possible gap-graded soil

Sieve Analysis Example

The results of a dry-sieving test are given together with the grading analysis and grading curve. Note carefully how the tabulated results are set out and calculated. The grading curve has been plotted on special semi-logarithmic paper; you can also do this analysis using a spreadsheet.

Sieve Mesh Size (mm)	Mass Retained (g)	Percentage Retained	Percentage Finer (passing)
14.0	0	0	100.0
10.0	3.5	1.2	98.8
6.3	7.6	2.6	86.2
5.0	7.0	2.4	93.8
3.35	14.3	4.9	88.9
2.0	21.1	7.2	81.7
1.18	56.7	19.4	62.3
0.600	73.4	25.1	37.2
0.425	22.2	7.6	29.6
0.300	26.9	9.2	20.4
0.212	18.4	6.3	14.1
0.150	15.2	5.2	8.9
0.063	17.5	6.0	2.9
Pan	8.5	2.9	
Total	292.3	100.0	

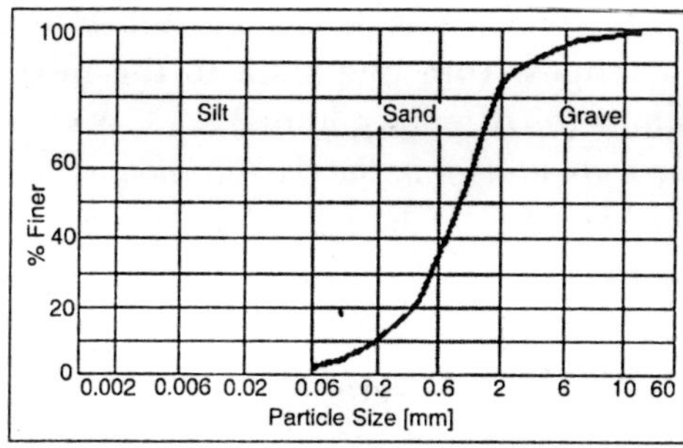

Fig. 1.18

The soil comprises: 18% gravel, 45% coarse sand, 24% medium sand, 10% fine sand, 3% silt, and is classified therefore as: a well-graded gravelly Sand

FINE SOILS

- Consistency limits and plasticity
- Plasticity index
- The plasticity chart and classification
- Activity

In the case of fine soils (*e.g.* Clays and Silts), it is the *shape* of the particles rather than their size that has the greater influence on engineering properties. Clay soils have flaky particles to which water adheres, thus imparting the property of plasticity.

Consistency Limits and Plasticity

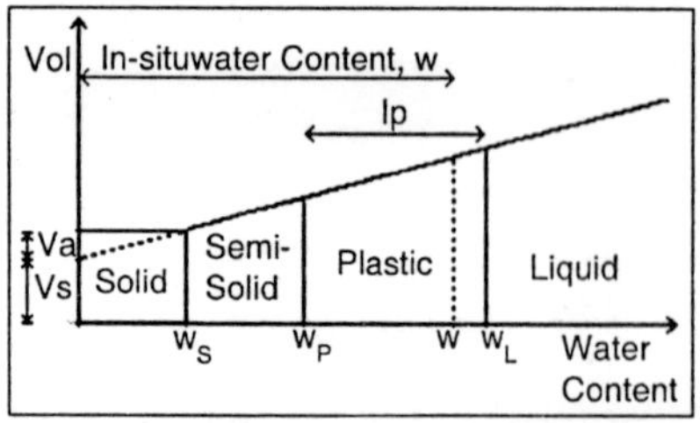

Fig. 1.19

Consistency varies with the water content of the soil. The consistency of a soil can range from (dry) *solid* to *semi-solid* to *plastic* to *liquid* (wet). The water contents at which the consistency changes from one state to the next are called consistency limits (or Atterberg limits).

Two of these are utilised in the classification of fine soils:

1. *Liquid Limit (w_L):* Change of consistency from plastic to liquid
2. *Plastic Limit (w_P):* Change of consistency from brittle/ crumbly to plastic

Measures of liquid and plastic limit values can be obtained from laboratory tests.

Plasticity Index

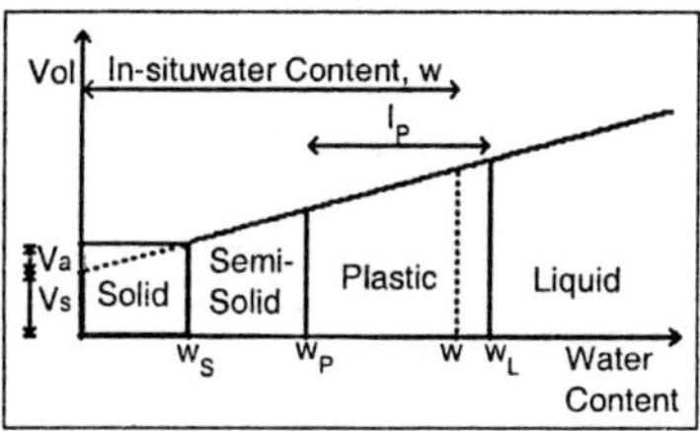

Fig. 1.20

The consistency of most soils in the ground will be plastic or semi-solid. Soil strength and stiffness behaviour are related to the range of plastic consistency. The range of water content over which a soil has a plastic consistency is termed the Plasticity Index (I_P or PI).

I_P = Liquid limit–plastic limit

= $w_L - w_P$

The Plasticity Chart and Classification

In the BSCS fine soils are divided into *ten* classes based on their measured plasticity index and liquid limit values: CLAYS are distinguished from SILTS, and five divisions of plasticity are defined:

Table. A Plasticity Chart is Provided to Aid Classification.

Low plasticity	w_L = < 35%
Intermediate plasticity	w_L = 35-50%
High plasticity	w_L = 50-70%
Very high plasticity	w_L = 70-90%
Extremely high plasticity	w_L = > 90%

Activity

So-called 'clay' soils are not 100% clay. The proportion of clay mineral flakes (< 2 μm size) in a fine soil affects its current state, particularly its tendency to swell and shrink with changes in water content. The degree of plasticity related to the clay content is called the activity of the soil.

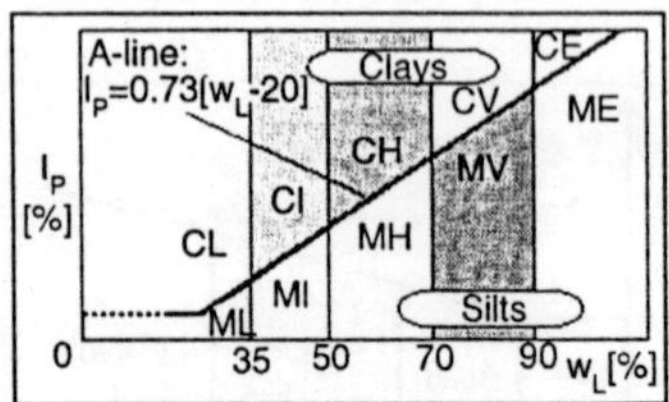

Fig. 1.21

Activity = I_P/ (% clay particles) Some typical values are:

Mineral	Activity	Soil	Activity
Muscovite	0.25	Kaolin clay	0.4-0.5
Kaolinite	0.40	Glacial clay and loess	0.5-0.75
Illite	0.90	Most British clays	0.75-1.25
Montmorillonite	> 1.25	Organic estuarine clay	> 1.25

SPECIFIC GRAVITY

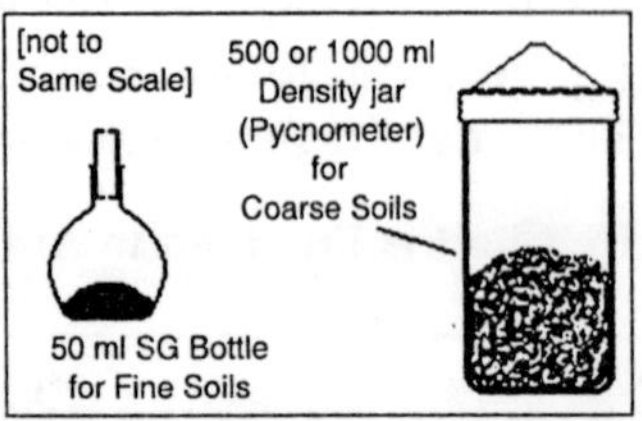

Fig. 1.22

Specific gravity (G_s) is a property of the mineral or rock material forming soil grains. It is defined as

$$G_S = \frac{\text{Mass of a soil grain}}{\text{Mass of an equal volume of water}}$$

Method of Measurement

For fine soils a 50 ml density bottle may be used; for coarse soils a 500 ml or 1000 ml jar. The jar is weighed empty (M_1). A quantity of dry soil is placed in the jar and the jar weighed (M_2).

The jar is filled with water, air removed by stirring, and weighed again (M_3). The jar is emptied, cleaned and refilled with water-and weighed again (M_4).

$$G_S = \frac{\text{Mass of soil}}{\text{Mass of water displaced by soil}}$$

$$= \frac{M_2 - M_1}{(M_4 - M_1) - (M_3 - M_2)}$$

[The range of G_s for common soils is 2.64 to 2.72]

VOLUME-WEIGHT PROPERTIES

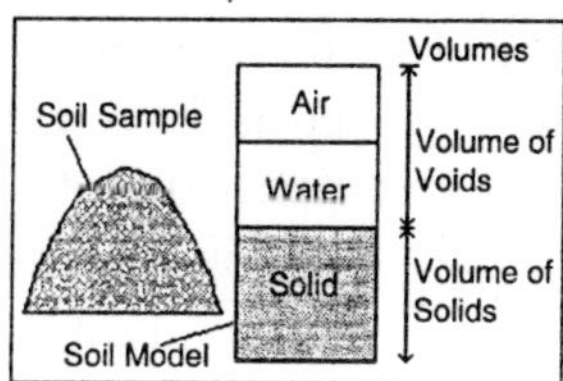

Fig. 1.23

The volume-weight properties of a soil define its state. Measures of the amount of void space, amount of water and the weight of a unit volume of soil are required in engineering analysis and design.

Soil comprises three constituent phases:

1. *Solid:* Rock fragments, mineral grains or flakes, organic matter.
2. *Liquid:* Water, with some dissolved compounds (*e.g.* salts).
3. *Gas:* Air or water vapour.

In natural soils the three phases are intermixed. To aid analysis it is convenient to consider a soil model in which the three phases are seen as separate, but still in their correct proportions.

VOLUMES OF SOLID, WATER AND AIR: THE SOIL MODEL

- Degree of saturation
- Air-voids content

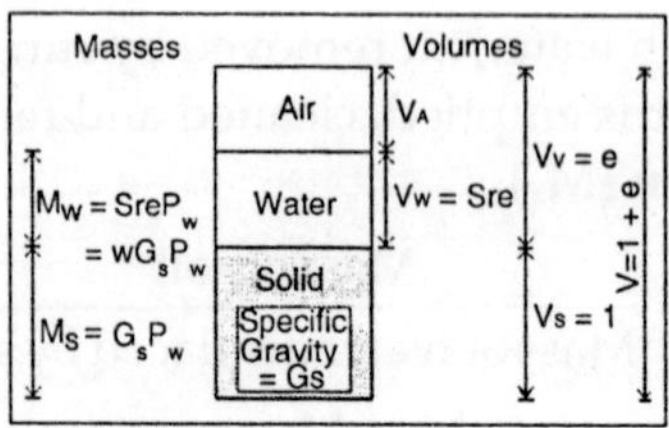

Fig. 1.24

The soil model is given dimensional values for the solid, water and air components: Total volume,

$$V = V_s + V_w + V_a$$

Since the amounts of both water and air are variable, the volume of solids present is taken as the reference quantity. Thus, the following relational volumetric quantities may be defined:

$$\text{Void ratio}, e = \frac{\text{Volume of voids}}{\text{Volume of solids}} = \frac{V_v}{V}$$

$$\text{Porosity}, n = \frac{\text{Volume of voids}}{\text{Total volume}} = \frac{V_v}{V}$$

$$\text{Specific volume}, v = 1 + e = \frac{V}{V_s}$$

Note also that:

$$n = e/\,(1 + e)$$
$$e = n/\,(1 - n)$$
$$v = 1/\,(1 - n)$$

Typical void ratios might be 0.3 (*e.g.* for a dense, well graded granular soil) or 1.5 (*e.g.* for a soft clay).

Degree of Saturation

The volume of water in a soil can only vary between zero (*i.e.* a dry soil) and the volume of voids; this can be expressed as a ratio:

$$\text{Degree of saturation}, S_r = \frac{\text{Volume of water}}{\text{Volume of voids}} = \frac{V_w}{V_v}$$

For a perfectly dry soil:

- $S_r = 0$

For a saturated soil:

- $S_r = 1$

Note: In clay soils as the amount water increases the volume and therefore the volume of voids will also increase, and so the degree of saturation may remain at $S_r = 1$ while the actual volume of water is increasing.

Air-Voids Content

The air-voids volume, V_a, is that part of the void space not occupied by water.

$$\begin{aligned} V_a &= V_v - V_w \\ &= e - e.S_r \\ &= e.(1 - S_r) \end{aligned}$$

Air-voids content, A_v

$$\begin{aligned} A_v &= \text{(Air-voids volume)/(Total volume)} \\ &= V_a / V \\ &= e.(1 - S_r) / (1 + e) \\ &= n.(1 - S_r) \end{aligned}$$

For a perfectly *dry* soil:

$$A_v = n$$

For a *saturated* soil:

$$A_v = 0$$

MASSES OF SOLID AND WATER: WATER CONTENT

The mass of air may be ignored. The mass of solid particles is usually expressed in terms of their particle density or grain specific gravity.

Grain Specific Gravity

$$G_s = \frac{\text{Mass of a soil grain}}{\text{Mass of an equal volume of water}}$$

Hence the mass of solid particles in a soil,

$$M_s = V_s.G_s.r_w$$

(r_w = density of water = 1.00Mg/m^3)

[Range of G_s for common soils: 2.64–2.72]

Particle Density

r_s = Mass per unit volume of particles

$= G_s.r_w$

The ratio of the mass of water present to the mass of solid particles is called the *water* content, or sometimes the moisture content.

$$\text{Water content, w} = \frac{\text{Mass of water}}{\text{Mass of solids}} = \frac{M_w}{M_s}$$

From the soil model it can be seen that,

$$w = (S_r.e.r_w)/\ (G_s.r_w)$$

Giving the useful relationship:

$$w.G_s = S_r.e$$

DENSITIES AND UNIT WEIGHTS

Density is a measure of the quantity of *mass* in a unit volume of material. Unit weight is a measure of the *weight* of a unit volume of material. There are two basic measures of density or unit weight applied to soils: Dry density is a measure of the amount of *solid* particles per unit volume. Bulk density is a measure of the amount of *solid* + *water* per unit volume.

$$\text{Dry density}, \rho_d = \frac{\text{Mass of solids}}{\text{Total volume}} = \frac{M_s}{V} = \frac{G_s\rho_w}{1+e}$$

$$\text{Bulk density}, \rho = \frac{\text{Total mass}}{\text{Total volume}} = \frac{M_s + M_w}{V} = \frac{G_s\rho_w + S_r e\rho_w}{1+e}$$

The preferred units of density are:

Mg/m³, kg/m³ or g/ml.

The corresponding *unit weights* are:

$$\text{Dry unit weight}, \gamma_d = \frac{\text{Dry weight}}{\text{Total volume}} = \frac{W_s}{V} = \frac{G_s\gamma_w}{1+e} = 9.81\rho d$$

$$\text{Unit weight}, \gamma = \frac{\text{Total weight}}{\text{Total volume}} = \frac{W_s + W_w}{V} = \frac{G_s\gamma_w + S_r e\gamma_w}{1+e} = 9.81\rho$$

Also, it can be shown that,

$$r = r_d(1 + w) \text{ and}$$

$$g = g_d(1 + w)$$

LABORATORY MEASUREMENTS

- Water content
- Unit weight

It is important to quantify the state of a soil immediately it is received in the testing laboratory and just prior to commencing other tests (*e.g.* shear tests, compression tests, etc.).The water content and unit weight are particularly important, since these could change during transportation and storage. Some physical state properties are calculated following the practical measurement of others; *e.g.* void ratio from porosity, dry unit weight from unit weight and water content.

Water Content

The most usual method of determining the water content of soil is to weigh a small representative specimen, drying it to constant weight and then weighing it again. Drying can be carried out using an electric oven set at 104–105° Celsius or using a microwave oven.

Example: A sample of soil was placed in a tin container and weighed, after which it was dried in an oven and then weighed again. Calculate the water content of the soil.

Weight of tin empty	= 16.16 g
Weight of tin + moist soil	= 37.82 g
Weight of tin + dry soil	= 34.68 g
Water content, w	= (Mass of water)/(Mass of dry soil)
	= (37.82–34.68)/ (34.68–16.16)
	= 0.169
Percentage water content	= 16.9%

Unit Weight

- *Clay soils:* Specimens are usually prepared in the form of regular geometric shapes, (*e.g.* prisms, cylinders) of which the volume is easily computed.
- *Sands and gravels:* Specimens have to be placed in a container to determine volume (*e.g.* a cylindrical can).

Example

A soil specimen had a volume of 89.13 ml, a mass before drying of 174.45 g and after drying of 158.73 g; the water content was 9.9%. Determine the bulk and dry densities and unit weights.

Bulk Density	
ρ	= (Mass of specimen)/(Volume of specimen)
	= 174.45/ 89.13 g/ml
	= 1.957 Mg/m³
[1 g/ml = 1 Mg/m³]	
Unit Weight	
γ	= 9.81m/s² × r Mg/m³
	= 19.20 kN/m³
Dry density	
ρ_d	= (Mass after drying)/(volume)
	= 158.73/ 89.13
	= 1.781 Mg/m³
ρ_d	= r/ (1 + w)
	= 1.957/ (1+0.099)
	= 1.781 Mg/m³
Dry unit weight}	
γ_d	= g/ (1 + w)
	= 19.20/ (1+0.099)
	= 17.47 kN/m³

FIELD MEASUREMENTS

Measurements taken in the field are mostly to determine density/unit weight. The most common application is the determination of the density of rolled and compacted fill, *e.g.* in road bases, embankments, etc.

Note: These methods are covered in detail by BS1377. You should understand the general principle that density is calculated from the mass and volume of a sample. How a sample of known volume is obtained depends on the nature of the soil. You are not expected to remember the details of each method

The Core Cutter Method

This method is suitable for soft fine grained soils.

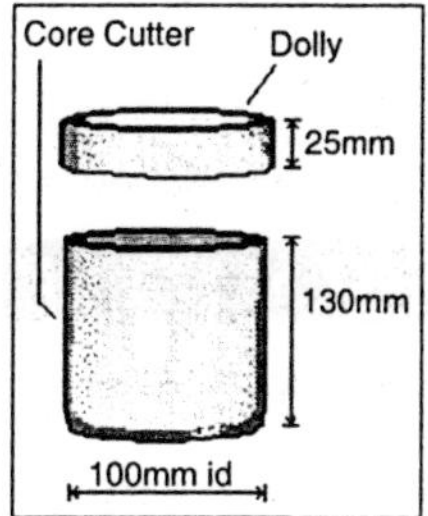

Fig. 1.25

A steel cylinder is driven into the ground, dug out and the soil shaved off level. The mass of soil is found by weighing and deducting the mass of the cylinder. Small samples are taken from both ends and the water content determined.

The Sand-Pouring Cylinder Method

This method is suitable for stony soils

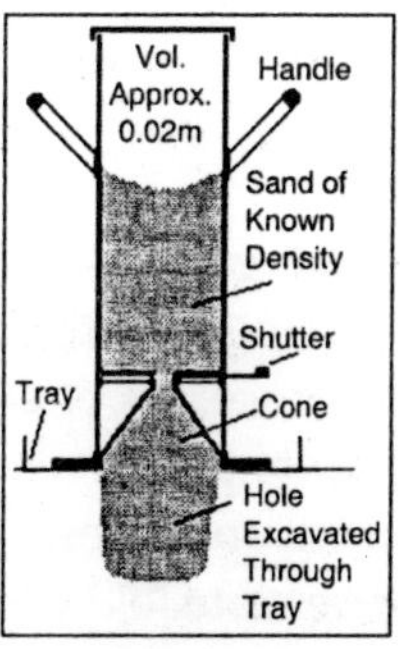

Fig. 1.26

Using a special tray with a hole in the centre, a hole is formed in the soil and the mass of soil removed is weighed. The volume of the hole is calculated from the mass of clean dry running sand required to fill the hole. The sand-pouring cylinder is used to fill the hole in a controlled manner. The mass of sand required to fill the hole is equal to the difference in the weight of the cylinder before and after filling the hole, less an allowance for the sand left in the cone above the hole as shown in figure 1.26.

Bulk Density

ρ = (Mass of soil)/(Volume of core cutter or hole)

CURRENT STATE OF SOIL

The state of soil is essentially the closeness of packing of the grains in the range:

Closely packed	→	Loosely packed
Dense	→	Loose
Low water content	→	High water content
Strong and stiff	→	Weak and soft

The important indicators of the current state of a soil are:

- *Current Stresses:* Vertical and horizontal effective stresses
- *Current Water Content:* Effecting strength and stiffness in fine soils
- *Liquidity Index:* Indicates state in fine soils
- *Density Index:* Indicates state of compaction in coarse soils
- *History of Loading and Unloading:* Degree of overconsolidation

Engineering operations (*e.g.* excavation, loading, unloading, compaction, etc.) on soil bring about changes in its state. Its initial state is the result of processes of erosion and deposition. It is possible for the engineer to predict changes that could result from a proposed engineering operation: changes from the soil's *current state* to a new future state.

SOIL HISTORY: DEPOSITION AND EROSION

Original Deposition

Most soils are formed in layers or lenses by deposition from moving water, ice or wind. One-dimensional compression occurs as overlying layers are added. Vertical and horizontal stresses increase with deposition.

Erosion

Erosion causes unloading; stresses decrease; some vertical expansion occurs. Plastic strain has occurred; the soil remains compressed, *i.e.* overconsolidated.

Subsequent Changes

Subsequent changes may occur in the depositional environment: further loading/unloading due to glaciation, land movement, engineering; and ageing processes.

SOIL HISTORY: AGEING

The term *ageing* includes processes that occur with time, except loading and unloading. Ageing processes are independent of changes in loading.

Vibration and Compaction

Coarse soils can be made more dense by vibration or compaction at essentially constant effective stress

Creep

Fine soils creep and continue to compress and distort at constant effective stress after primary consolidation is complete.

Cementing and Bonding

Intergranular cementing and bonding occurs due to deposition of minerals from groundwater, *e.g.* calcium carbonate; disturbance due to excavation fractures the bonding and reduces strength.

Weathering

Physical and chemical changes take place in soils near the ground surface due to the influence of changes in rainfall and temperature.

Changes in Salinity

Changes in the salinity of groundwater are due to changes in relative sea and land levels, thus soil originally deposited in sea water may later have fresh water in its pores, such soils may be prone to sudden collapse.

DENSITY INDEX (RELATIVE DENSITY)

The void ratio of coarse soils (sands and gravels) varies with the state of packing between the loosest practical state in which it can exist and the densest. Some engineering properties are affected by this,

Shear Strength

Seepage

The flow of water through soil.

Seepage Force (J)

The force transmitted to a body of soil due to the seepage of groundwater. $J = i\,\gamma_w\,V$

Seepage Pressure (j)

The seepage force per unit volume $j = i\,\gamma_w$

Seepage Velocity (v_s)

The average velocity at which groundwater flows through the pores; the ratio of the volume flow rate to the average area of voids in a cross-section.

$$v_s = q/\,A_v$$

Sensitivity (S_t)

$$S_t = \frac{\text{Undisturbed undrained strength}}{\text{Remoulded undrained strength}}$$

A measure of the change in strength of f clays upon disturbance: For ordinary clays $S_t = 1$ to 4, sensitive clays 4 to 8, 'quick' clays 16–100.

Settlement (s, ρ)

The downward movement of ground or ground surface; the downward movement of a foundation.

Shape Factors ($s_c\, s_q\, s_g$)

Factors used in a general equation giving ultimate bearing capacity which provide adjustment relating to the shape (*e.g.* strip, square, circle)

Shear Modulus (G')

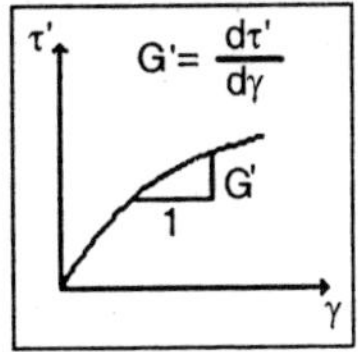

Fig. 1.27

The ratio of the change in shear stress to the resulting change in shear strain.

Shear Strain (γ)

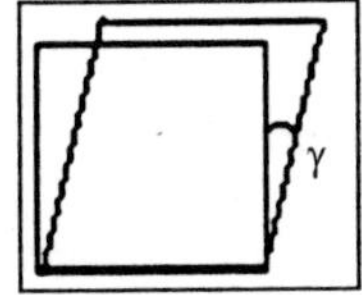

Fig. 1.28

The angular distortion or change in shape of a body of material.

Shear Strength (τ_f)

The maximum shear stress which a material can sustain under a given set of conditions. In soil mechanics it is necessary to refer shear strength to the strain at which the strength is measured.

- *Critical shear strengt*: $\tau_c = c'_c + \sigma' \tan\phi'_c$
- *Peak shear strength*: $\tau_p = c'_p + \sigma' \tan\phi'_p$
- *Undrained shear strength*: $\tau_c = s_u$
- *Residual shear strength*: $\tau_r = c'_r + \sigma' \tan\phi'_r$

Shear Stress (τ)

The force per unit area acting tangentially to a given plane or surface. Units: kPa.

Short-term Conditions

Conditions in the ground when it is undrained; loading or unloading will cause changes in pore pressures, but not immediately in volume; these excess pore pressures dissipate with time due to consolidation.

Shrinkage Limit (w_s)

(Also SL) The water content below which further reduction in water content causes no further reduction in volume.

Skin Friction Stress (f_s)

The shear stress on the shaft of a pile of a pile or caisson or cone penetrometer. *Note:* Although known as 'skin friction', the shear stress may not vary with normal stress.

Soil Suction

Negative pore pressure created by capillary attraction in fine soils and in unsaturated soils.

Specific Gravity (G_s)

The ratio of the mass of a body or a substance to the mass of an equal volume of water; the ratio of the density of a body or a substance to that of water.

$$G_s = \frac{\text{Mass of body}}{\text{Mass of the same volume of water}}$$

Specific Surface (S_s)

The total surface area of all particles in a unit mass of soil. Units: m^2/g.

Specific Volume (v)

The total volume of a quantity of soil containing a unit volume of soil grains.

$$v = 1 + e$$

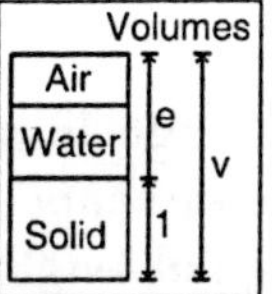

Fig. 1.29

State Boundary Surface

The state boundary surface (SBS) is the boundary (usually in q′:p′:v space) to all possible stress-volume states of a soil.

If a soil with a state on the state boundary surface is unloaded, the subsequent state will be inside the SBS, upon re-loading the subsequent state will move back onto (but not beyond) the SBS. In other words, stable states cannot exist outside the SBS.

State Parameter (S_s, S_v)

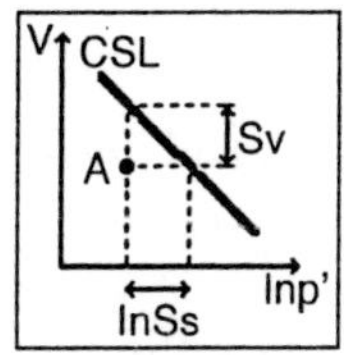

Fig 1.30

A measure of the distance between the current state (A) and the critical state; expressed as a ratio of stresses or as a difference of specific volumes,

$$S_s = p'_A / p'_c$$

$$S_v = v_A - v_c$$

Steady State Pore Pressure (u_o)

The pore pressure at equilibrium when all excess pore pressures have fully dissipated.

Stiffness

Susceptibility to distortion or volume change under load.

Strain

A measure of the change in size or shape of a body, relative to its original size or shape. (Direct strain is the ratio of change in length to original length; shear strain is the angle of distortion; volumetric strain is the ratio of change in volume to original volume.)

Stream Function (Ψ)

A function introduced in the solution of the Laplace equations defining two-dimensional seepage flow; the flow-quantity interval ($\Delta\theta$) between stream lines (flow lines) can be stated as $\Delta\Psi = \Delta\theta$.

Stress

The intensity of force per unit area; normal stress is applied perpendicularly to a surface or plane, shear stress is applied tangentially to a surface or plane.

Stress History

The past history of loading and unloading associated with a soil.

Summation (Σ)

The symbol Σ when placed in front of a quantity indicates that the quantity is a sum or total.

Swelling/Recompression Index (C_s, k)

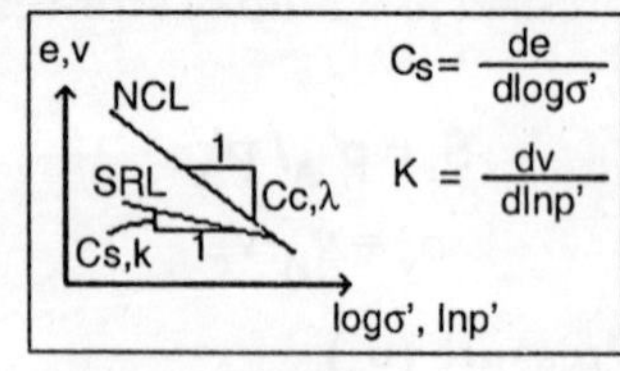

Fig. 1.31

The slope of the swelling (unloading) and recompression (reloading)line. $C_s = 2.303$ k where k = slope of v:ln σ′ swelling/ recompression line.

Compressibility

Capillary Rise (h_c)

The height to which water will rise above the water table due to capillarity.

Clay

Soil particles less than 0.002 mm in size.

Coarse-grained Soils

Soils with less than 35% by mass of grains less than 0.06mm in size:

- *Coarse soil*: Grain size is predominantly 0.06–60 mm
- *Very coarse soil*: Grain size is predominantly greater than 60mm

Coefficient of Active Earth Pressure (K_a)

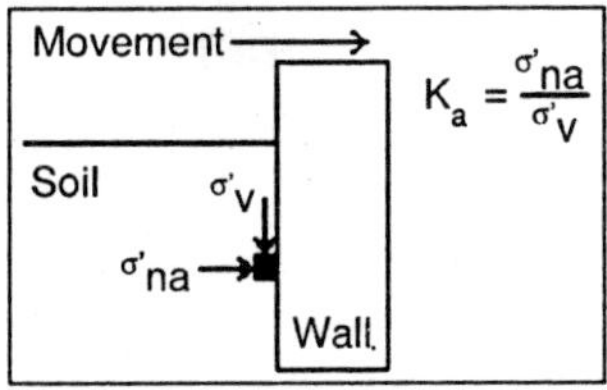

Fig. 1.32

The ratio of the minimum horizontal effective stress to the vertical effective stress at a point in a soil mass retained by a surface as the surface moves away from the soil.

Coefficient of Consolidation (c_v and c_h)

$$C_v \frac{\partial^2 \bar{u}}{\partial z^2} = \frac{\partial \bar{u}}{\partial t}$$

A measure of the rate of change of volume during primary consolidation; referred to either vertical drainage (c_v) or horizontal drainage (c_h). Units: m^2/s

Coefficient of Curvature of Grading (C_z)

(Also curvature coefficient) A measure of the shape of a grading curve: $C_z = d_{30}^2 / (d_{60} \times d_{10})$.

Coefficient of Earth Pressure at Rest (K_o)

The ratio of horizontal effective stress to vertical effective stress at a point in a soil mass loaded in conditions of zero horizontal strain.

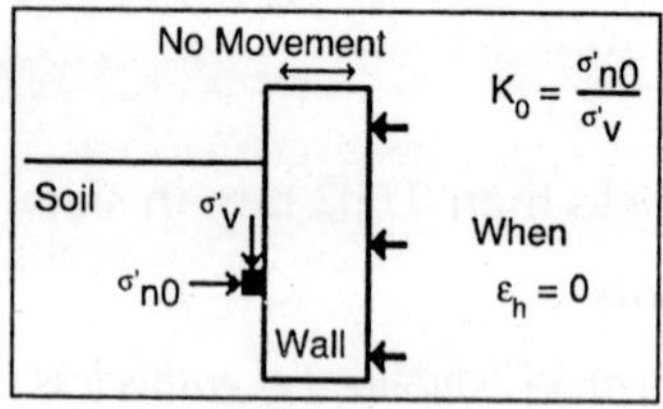

Fig. 1.33

Coefficient of Friction (μ)

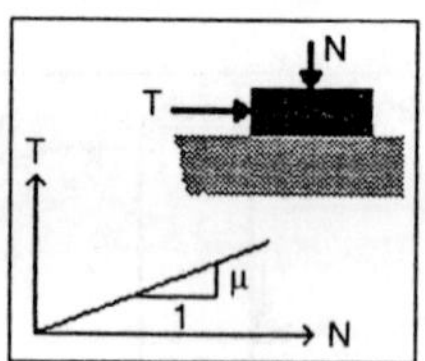

Fig. 1.34

The ratio between the tangential force (T) required to cause a body to slide along a plane and the normal force (N) between the body and the plane: $T = \mu N$

Along a wall or foundation surface:

$$\mu = \tan \delta$$

where δ = angle of wall friction,

Coefficient of Passive Earth Pressure (K_p)

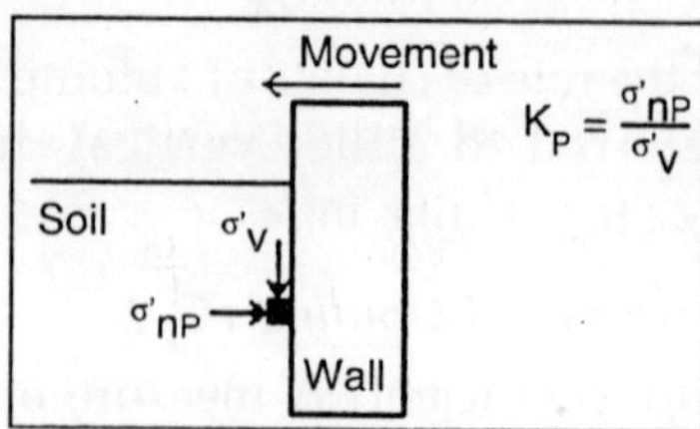

Fig. 1.35

The ratio of the maximum horizontal effective stress to the vertical effective stress at a point in a soil mass retained by a surface as the surface moves towards from the soil.

Coefficient of Permeability (k)

The constant average discharge velocity (v) of water passing through soil when the hydraulic gradient (i) is 1.0; defined by Darcy's law: v = k.i

Coefficient of Secondary Consolidation (C_a)

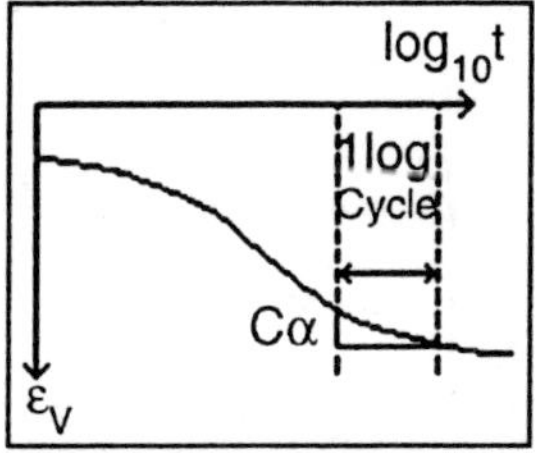

Fig. 1.36

The change in volumetric strain per $\log_{10}$ cycle of time after primary consolidation is complete.

Coefficient of Uniformity of Grading (C_u)

(Also uniformity coefficient) A measure of the slope of a grading curve and therefore the uniformity of the soil inparticle size analyses: $C_u = d_{60}/\ d_{10}$

Coefficient of Volume Compressibility (m_v)

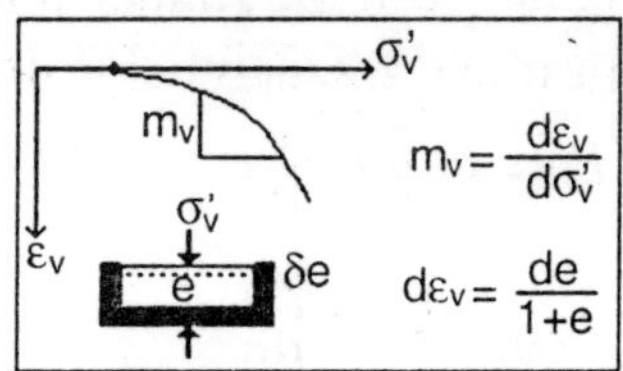

Fig. 1.37

The change in volumetric strain per unit volume per unit change in effective stress in one-dimensional compression. (Units m^2/MN) Compressibility is the reciprocal of stiffness.

Cohesion (c′)

Apparent cohesion (c′) is the intercept on the shear stress axis of a straight-line Mohr-Coulomb envelope. For critical states and residual states, $c'_c = c'_r = 0$ in most cases. For peak states, a curved strength envelope also passes through the origin, but a straight-line fitted to a small number of results is often extrapolated to give an intercept $c'_p > 0$. (In physics, cohesion is described as 'the force that holds together molecules or like particles within a substance'.)

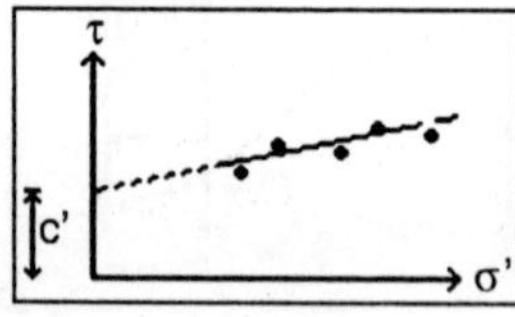

Fig. 1.38

Compaction

Volume change in soil in which air is expelled, but with the water content remaining constant. Compaction may occur due to vibration in loose sands and gravels, and in fill due to self-weight. In soil constructions, compaction is achieved by rolling, tamping or vibrating.

Compatibility

The relationship between the strains in a deforming body so that no holes appear and no material is destroyed.

Compressibility of Pore Fluid (C_v)

$$C_v = \frac{d\varepsilon_v}{d\sigma}$$

The ratio of the change in volumetric strain to the change in isotropic stress. For a saturated soil, $C_v = \infty$

Compression Index (C_c)

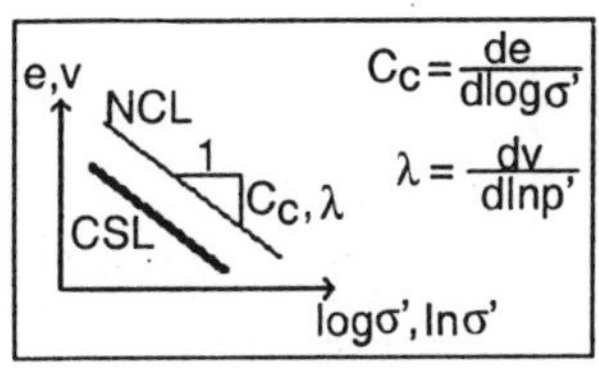

Fig. 1.39

The slope of the normal compression line (NCL) and critical state line (CSL).

Cone Resistance (q_c)

The resistance force divided by the end area of the cone tip, measured during the cone penetration test.

Cone Penetration Test

A penetration test in which a sleeved cone (diameter 35.7mm, end area 1000mm^2, apex angle 60°) is pushed into the ground at a rate of 20 mm/min and the force required measured. The magnitude of this force divided by the end area is called the *cone penetration resistance* (q_c). The cone and the sleeve can be advanced separately and so the frictional resistance along the sleeve (1000mm^2 area) or *local side friction* (f_s) can also be measured.

Confined Aquifer

A *confined* aquifer is contained between two strata of low permeability; in flow analyses, a confined aquifer is often assumed to be saturated throughout its depth.

Consistency Index (I_c)

A measure of the relationship between the current water content and the consistency limits, similar in form to the density index, but expressed in terms of water content.

$$I_c = (w_L - w)/I_P$$

- At liquid limit, $I_c = 0$
- At plastic limit, $I_c = 1.0$

Consistency Limits

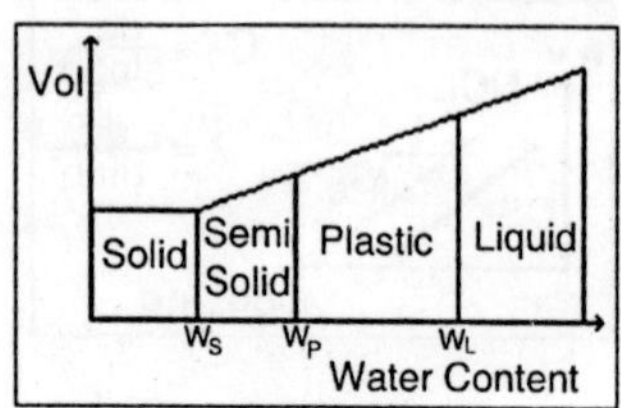

Fig. 1.40

(Also called Atterberg limits) Measures of water content corresponding to changes in physical state of a soil: liquid limit (w_L) = w/c at the change from liquid to plastic plastic limit (w_P) = w/c at the change from plastic to semi-solid shrinkage limit (w_s) = w/c below which no further shrinkage upon drying occurs

Consolidation

Volume change due to dissipation of excess pore pressure, usually with constant total stress.

Consolidation Settlement (s_c, ρ_c)

The settlement of a foundation due to consolidation.

Coulomb's Equation

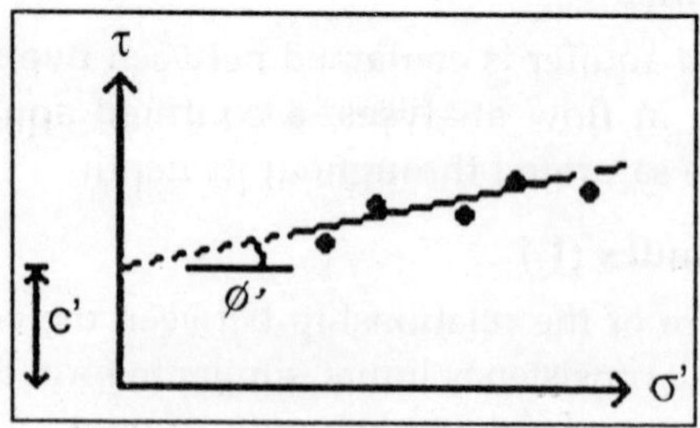

Fig. 1.41

(After Charles Augustin Coulomb, 1736–1806) An

equation relating the shear strength of soil to the normal effective stress on a failure plane.

$$\tau_f = c' + \sigma' \tan\phi' = c' + (\sigma - u) \tan\phi'$$

Creep

Deformation or volume change which occurs in soil at constant effective stress progressing with time.

Critical Circle

In slope stability analyses the slip circle corresponding to the lowest factor of safety.

Critical Ground Slope Angle (i_c)

The ground slope angle that corresponds to a slope-stability factor of safety of 1.0.

Critical Height (H_c)

The height of a slope (*e.g.* embankment, cutting, trench) for which the factor of safety against collapse is 1.0.

Critical Hydraulic Gradient (i_c)

The hydraulic gradient at which effective stresses becomes zero; with upward seepage, sand may become quicksand.

Critical Shear Strength (τ_c)

The shear stress developed along a slip surface during shearing at constant volume; also known as ultimate shear strength.

Critical State

The state of a soil in which it strains; critical states occur on the critical state line.

Note: The critical state is not the same as the residual state; at the critical state, soil particles continue to rotate and the flow is turbulent; at the residual state, in clay soils, flat particles become aligned to the slip plane and the flow is laminar.

Critical State Line (CSL)

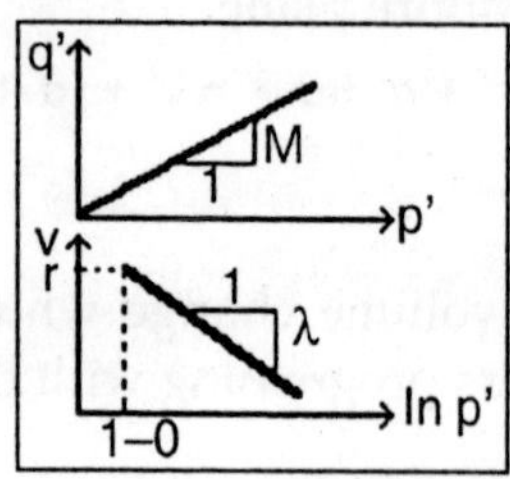

Fig. 1.42

The unique relationship at failure between deviator stress, average normal stress and volume (or shear stress, normal stress and void ratio) is defined by the critical state line.

$$q' = Mp'$$

$$v = \Gamma - \lambda \ln p'$$

$$\tau' = \sigma'_n \tan\phi'$$

$$e = e_G - C_c \log\sigma'_n$$

Critical Void Ratio (e_c)

The void ratio of a soil at which its volume remains constant during shearing. Note: critical void ratio depends on the mean stress.

Current State of Soil

The current state of a soil is described by its voids ratio (e) or specific volume (v), the current stress (σ' or p') and the overconsolidation ratio or yield stress ratio (R_y).

Ppermeability

The property which allows the flow of water through a soil.

pF Index

A measure of soil suction: pF = $\log_{10}$ (suction head in cm),*i.e.* for a suction of 100 kPa, pF = 2.0 [Range for soils is pF = 0 to 7]

pH Value

A measure of acidity or alkalinity of groundwater or soil water extract based on the hydrogen ion content:

- $pH = -\log_{10}$(hydrogen ion content)
- pH $<$ 7.0 indicates acidity.
- pH $>$ 7.0 indicates alkalinity.

Piezometer

An instrument used to measure *in situ* pore pressures; may be an open standpipe or an enclosed electronic pressure transducer.

Piezometric Surface

An imaginary surface corresponding to the hydrostatic water level of a confined body of groundwater; the notional level to which artesian pressure would raise water in a (real or imaginary) standpipe.

Pile Spacing (s)

The distance from centre to centre of piles in a group.

Plane Strain

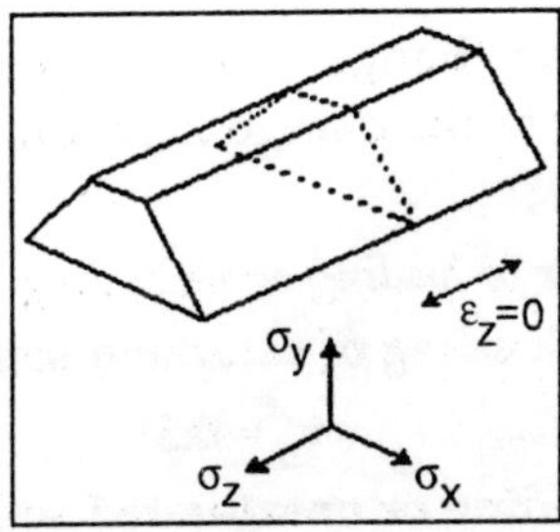

Fig. 1.43

A two-dimensional stress state, where the out-of-plane strain (*i.e.* the strain normal to the plane being considered, e_z) is zero. An example of a plane strain situation would be on a cross-section through a long structure being loaded in the x-y plane, such as an embankment dam.

Plastic Deformation

The flow or distortion resulting in a permanent and irrecoverable change in shape or volume.

Plastic Limit (w_p)

(Also PL) The moisture content above which a soil will have a plastic consistency, but below which it crumbles.

Plastic Strain

Deformation or strain that is not recovered upon unloading.

Plasticity

1. The property of a soil (or other material) which allows it to deform continuously.
2. Plasticity theory is used to calculate plastic (irreversible) deformations.

Plasticity Index (I_p)

(Also PI) The difference between the liquid limit and plastic limit. $I_p = w_L - w_P$

Poisson's Ratio (ν, ν')

The ratio of the change in strain perpendicular to the direction of loading to the change in strain caused in the same direction.

$\nu' = -d\varepsilon_y / d\varepsilon_x$ *(for to loading or unloading in the x-direction).*

For undrained loading of saturated soil,

$$\nu_u = 0.5$$

For drained loading or unsaturated soil,

$$\nu' = 0.2\text{–}0.5$$

Pore Air Pressure (u_a)

The pressure of air in a partially saturated soil; not necessarily the same as pore water pressure due to the surface tension on air-water interfaces within the voids.

Pore Pressure (u)

The pressure exerted by the fluid within the pores or voids in a porous material; in saturated soil the pore pressure is the pore water pressure.

Pore Pressure Coefficient (A)

The ratio of the change in pore pressure to the change in deviator stress, *e.g.* in an undrained triaxial test; the value of A varies with strain and the overconsolidation ratio.

Pore Pressure Coefficient (B)

The ratio of the change in pore pressure to the change in isotropic stress in undrained loading.

$$B = \frac{\Delta u}{\Delta p}$$

For saturated soils, B = 1.

Pore Pressure Force (U)

The resultant force due to pore pressure acting on a given area.

Pore Pressure Ratio (r_u)

$$r_u = \frac{u}{\sigma_v}$$

At a given point in a body of soil, the ratio of the porewater pressure to the vertical overburden pressure.

Pore Water Pressure (u_w)

Porosity (n)

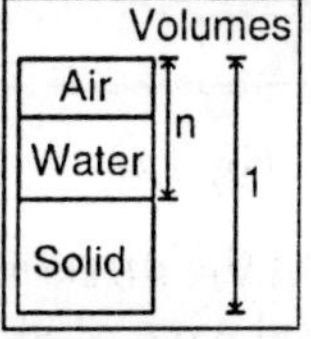

Fig. 1.44

The ratio of void volume to total volume:

$$n = V_v / V$$

where,

V_v = Volume of voids
V = Total volume

Potential Function (Φ)

A function introduced in the solution of the Laplace equations defining two-dimensional seepage flow.

Pressure Head (h_w)

(Also head) The height of a column of water required to develop a given pressure u at a given point.

$$h_w = u / \gamma_w$$

Pressure in Tension Crack (p_w)

The horizontal pressure exerted in a slope or against a retaining wall due to hydrostatic water pressure in tension cracks.

Principal Axes

A set of orthogonal axes perpendicular to which the shear stresses and shear strains are zero and normal stresses and strains are referred to as principal stresses and principal strains.

Principal Strains ($\varepsilon_1, \varepsilon_2, \varepsilon_3$)

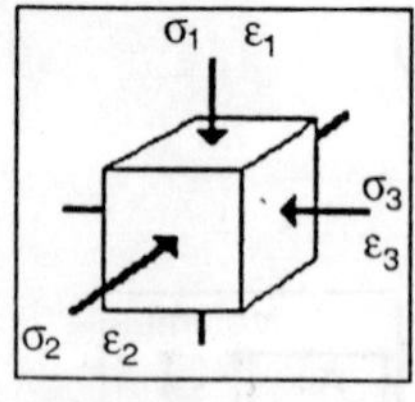

Fig. 1.45

The strains occurring in the directions of the principal axes of strain. Note: the principal axes of stress and strain may not coincide.

Principal Stresses (σ_1, σ_2, σ_3)

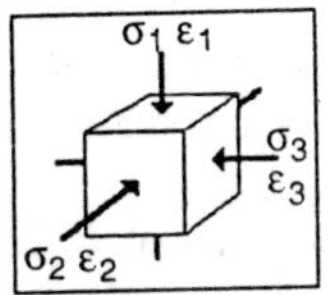

Fig. 1.46

Normal stresses acting in the direction of principal axes of stress. Note: the principal axes of stress and strain may not coincide. It is therefore useful to measure the *in situ* state and this can be done by comparing the *in situ* void ratio (e) with the minimum and maximum practical values (e_{min} and e_{max}) to give a density index I_D

$$I_D = \frac{e_{max} - e}{e_{max} - e_{min}}$$

e_{min} is determined with soil compacted densely in a metal mould
e_{max} is determined with soil poured loosely into a metal mould

Table. Density Index is Also known as Relative Density Relative States of Compaction are Defined:

Density index	State of compaction
0-15%	Very loose
15-35	Loose
35-65	Medium
65-85	Dense
85-100%	Very dense

LIQUIDITY INDEX

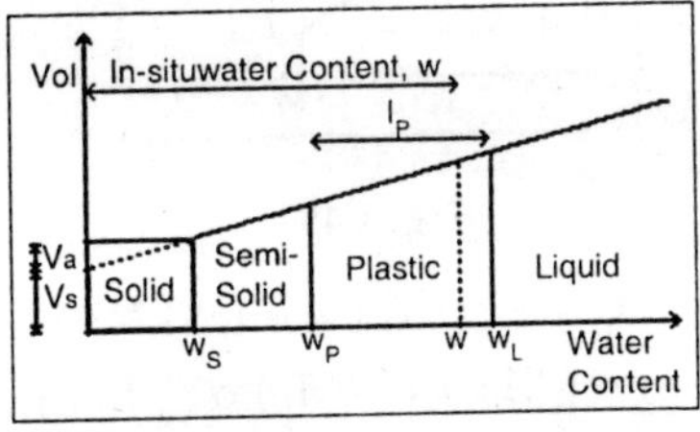

Fig. 1.47

In fine soils, especially clays, the current state is dependent on the water content with respect to the consistency limits (or Atterberg limits). The liquidity index (I_L or LI) provides a quantitative measure of the current state:

$$I_L = \frac{w - w_P}{I_P} = \frac{w - w_P}{w_L - w_P}$$

where,

w_P = plastic limit and

w_L = liquid limit

Significant values of I_L indicating the consistency of the soil are:

$I_L < 0 \Rightarrow$ semi-plastic solid or solid

$0 < I_L < 1 \Rightarrow$ plastic

$1 < I_L \Rightarrow$ liquid

PREDICTING STIFFNESS AND STRENGTH FROM INDEX PROPERTIES

Preliminary estimates of strength and stiffness can provide a useful basis for early design and feasibility studies, and also the planning of more detailed testing programmes. The following suggestions have been made; they are simple, but not necessarily reliable, and should be not be used in final design calculations.

Undrained Shear Strength

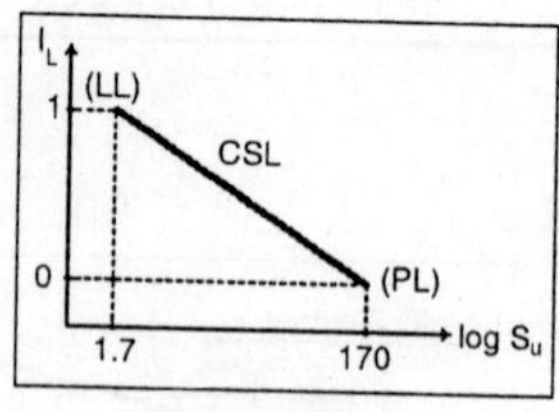

Fig. 1.48

$$s_u = 170 \exp(-4.6\ I_L)\ \text{kN/m}^2$$

$$s_u = (0.11 + 0.37\ I_P)\ \sigma'_{vo}\ \text{kN/m}^2$$

where σ'_{vo} = vertical effective stress in situ,

Stiffness

The slope of the critical state line may be estimated from:

$$\lambda = I_P.G_s / 461$$

The compressibility index may be estimated from:

$$C_c = \lambda \ In10 = I_P \ G_s / 200$$

(where P is in percentage units)

THE COMPRESSIBILITY INDEX MAY BECLASSIFICATION

BS 5930 Site Investigation recommends the terminology and a system for describing and classifying soils for engineering purposes. Without the use of a satisfactory system of description and classification, the description of materials found on a site would be meaningless or even misleading, and it would be difficult to apply experience to future projects.

BS DESCRIPTION SYSTEM

A recommended protocol for describing a soil deposit uses ninecharacteristics; these should be written in the following order:

- *Compactness*: *e.g.* loos , dense, slightly cemented
- *Bedding structure*: *e.g.* homogeneous or stratified; dip, orientation
- *Discontinuities*: spacing of beds, joints, fissures
- *Weathered state*: degree of weathering
- *Colour*: main body colour, mottling
- *Grading or consistency*: *e.g.* well-graded, poorly-graded; soft, firm, hard
- *Soil Name*: *e.g.* Gravel, Sand, Silt, Clay; plus silty, gravelly-, with-fines, etc. as appropriate
- *Soil class*: (BSCS) designation (for roads and airfields) *e.g.* SW = well-graded sand
- *Geological stratigraphic name*: (when known) *e.g.* London clay

Not all characteristics are necessarily applicable in every case.

Example:

- Loose homogeneous reddish-yellow poorly-graded medium SAND (SP), Flood plain alluvium

- Dense fissured unweathered greyish-blue firm CLAY. Oxford clay.

DEFINITIONS OF TERMS USED IN DESCRIPTION

A table is given in BS 5930 Site Investigation setting out a recommended field indentification and description system.

The following are some of the terms listed for use in soil descriptions:

- *Particle shape*: Angular, sub-angular, sub-rounded, rounded, flat, elongate
- *Compactness*: Loose, medium dense, dense (use a pick or driven peg, or density index)
- *Bedding structure*: Homogeneous, stratified, inter-stratified
- *Bedding spacing*: Massive(>2m), thickly bedded (2000-600 mm), medium bedded (600-200 mm), thinly bedded (200-60 mm), very thinly bedded (60-20 mm), laminated (20-6 mm), thinly laminated (<6 mm).
- Discontinuities: *i.e.* spacing of joints and fissure: very widely spaced(>2m), widely spaced (2000-600 mm), medium spaced (600-200 mm), closely spaced (200-60 mm), very closely spaced (60-20 mm), extremely closely spaced (<20 mm).
- *Colours*: Red, pink, yellow, brown, olive, green, blue, white, grey, black
- *Consistency*: Very soft (exudes between fingers), soft (easily mouldable), firm (strong finger pressure required), stiff (can be indented with fingers, but not moulded) very stiff (indented by sharp object), hard (difficult to indent).
- *Grading*: Well graded (wide size range), uniform (very narrow size range), poorly graded (narrow or uneven size range).
- *Composite soils*: In Sands and Gravels: slightly clayey or silty (<5%), clayey or silty (5-15%), very clayey or silty(>15%) In Clays And Silts: sandy or gravelly (35-65%)

Chapter 2

Mechanics of Soils

INTRODUCTION

Loads from foundations and walls apply stresses in the ground. Settlements are caused by strains in the ground. To analyse the conditions within a material under loading, we must consider the stress-strain behaviour.

The relationship between a strain and stress is termed stiffness. The maximum value of stress that may be sustained is termed strength.

ANALYSIS OF STRESS AND STRAIN

Stresses and strains occur in all directions and to do settlement and stability analyses it is often necessary to relate the stresses in a particular direction to those in other directions.

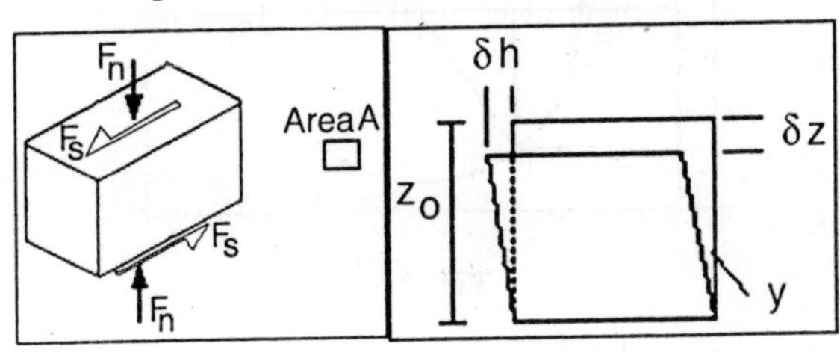

Fig. 2.1

- *Normal stress*: $\sigma = F_n / A$
- *Shear stress*: $\tau = F_s / A$
- *Normal strain*: $\varepsilon = \delta z / z_o$
- *Shear strain*: $\gamma = \delta h / z_o$

Note that compressive stresses and strains are positive, counter-clockwise shear stress and strain are positive, and that these are total stresses.

Mohr Circle Construction

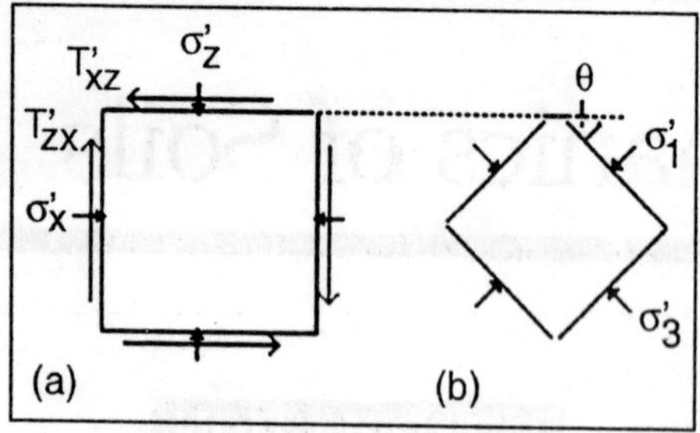

Fig. 2.5

Values of normal stress and shear stress must relate to a particular plane within an element of soil. In general, the stresses on another plane will be different. To visualise the stresses on all the possible planes, a graph called the Mohr circle is drawn by plotting a (normal stress, shear stress) point for a plane at every possible angle.

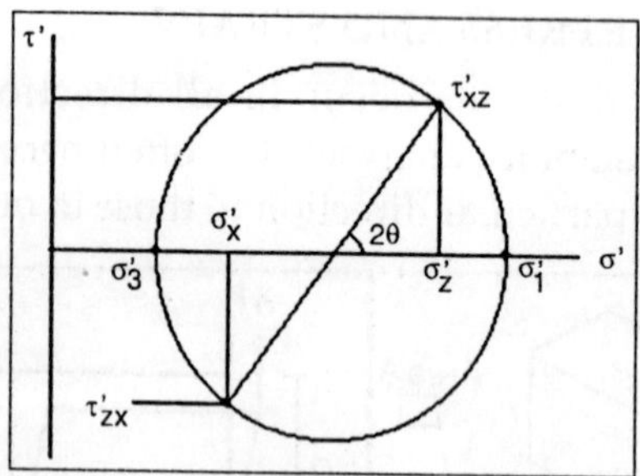

Fig. 2.6

There are special planes on which the shear stress is zero *(i.e. the circle crosses the normal stress axis)*, and the state of stress *(i.e. the circle)* can be described by the normal stresses acting on these planes; these are called the principal stresses σ'_1 and σ'_3.

Parameters For Stress and Strain

In common soil tests, cylindrical samples are used in which the axial and radial stresses and strains are principal stresses and strains. For analysis of test data, and to develop soil mechanics theories, it is usual to combine these into mean (or normal) components which influence volume changes, and deviator (or shearing) components which influence shape changes.

	Stress	Strain
Mean	$p' = (\sigma'_a + 2\sigma'_r)/3$	$e_v = \Delta V/V = (\varepsilon_a + 2\varepsilon_r)$
	$s' = \sigma'_a + \sigma'_r)/2$	$\varepsilon_n = (\varepsilon_a + \varepsilon_r)$
Deviator	$q' = (\sigma'_a - \sigma'_r)$	$e_s = 2(\varepsilon_a - \varepsilon_r)/3$
	$t' = (\sigma'_a - \sigma'_r)/2$	$\varepsilon_\gamma = (\varepsilon_a - \varepsilon_r)$

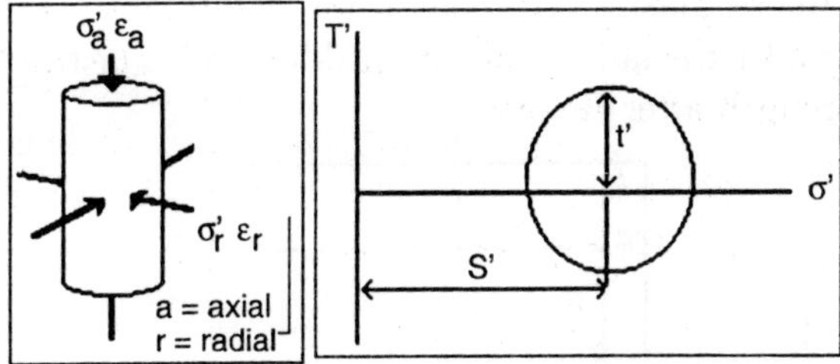

Fig. 2.7

In the Mohr circle construction t′ is the radius of the circle and s′ defines its centre.

Note: Total and effective stresses are related to pore pressure u:

$$p' = p - u$$

$$s' = s - u$$

$$q' = q$$

$$t' = t$$

Strength Back to Basic mechanics of soils:

- Types of failure
- Strength criteria

- Typical values of shear strength

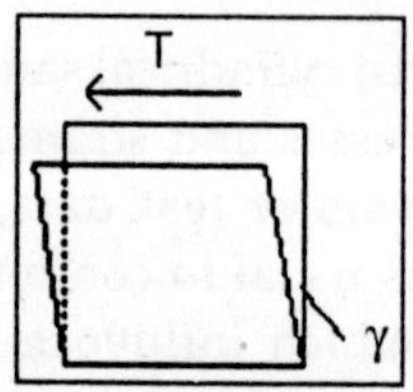

Fig. 2.8

The shear strength of a material is most simply described as the maximum shear stress it can sustain: When the shear stress t is increased, the shear strain γ increases; there will be a limiting condition at which the shear strain becomes very large and the material fails; the shear stress t_f is then the shear strength of the material. The simple type of failure shown here is associated with ductile or plastic materials. If the material is brittle (like a piece of chalk), the failure may be sudden and catastrophic with loss of strength after failure.

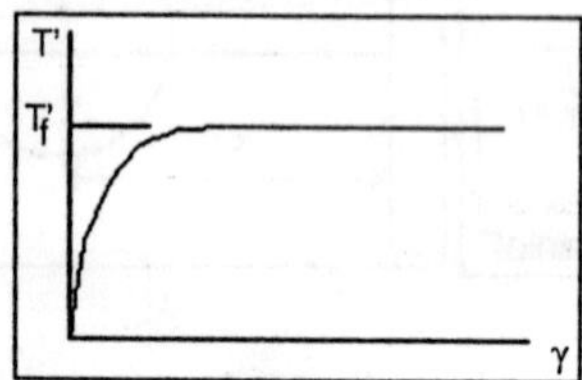

Fig. 2.9

TYPES OF FAILURE

Materials can 'fail' under different loading conditions. In each case, however, failure is associated with the limiting radius of the Mohr circle, *i.e.* the maximum shear stress. *The following common examples are shown in terms of total stresses*:

Shearing

- Shear strength = τ_f
- σ_{nf} = Normal stress at failure

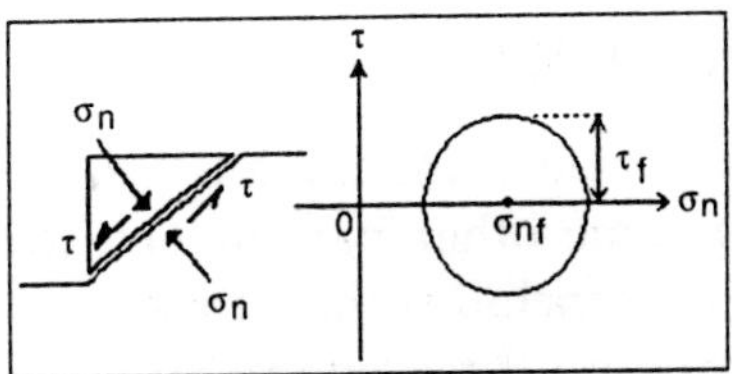

Fig. 2.10

Uniaxial Extension

- Tensile strength $\sigma_{tf} = 2\tau_f$

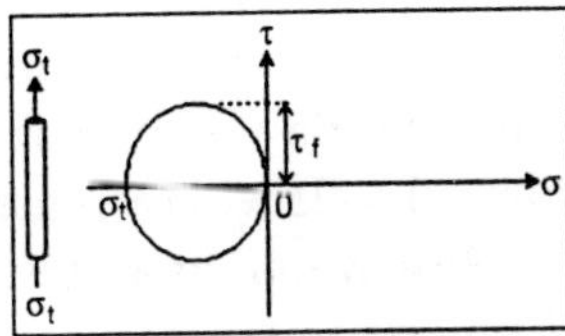

Fig. 2.11

Uniaxial Compression

- Compressive strength $\sigma_{cf} = 2\tau_f$

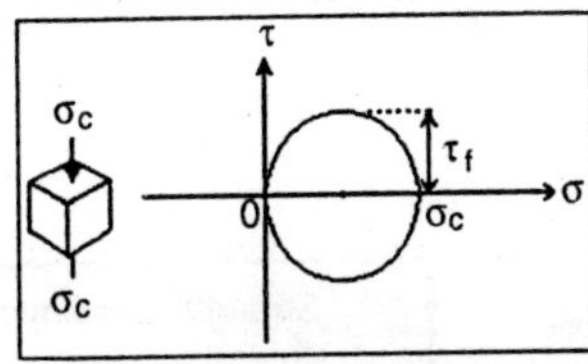

Fig. 2.12

Note:Water has no strength $\tau_f = 0$.

Hence vertical and horizontal stresses are equal and the Mohr circle becomes a point.

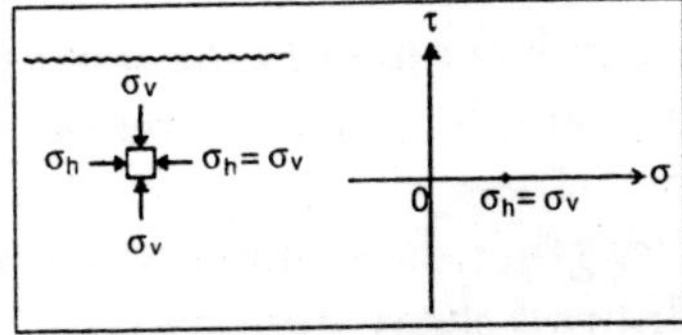

Fig. 2.13

STRENGTH CRITERIA

A strength criterion is a formula which relates the strength of a material to some other parameters: these are material parameters and may include other stresses.

For soils there are three important strength criteria: the correct criterion depends on the nature of the soil and on whether the loading is drained or undrained.

In General, course grained soils will "drain" very quickly (in engineering terms) following loading. Thefore development of excess pore pressure will not occur; volume change associated with increments of effective stress will control the behaviour and the Mohr-Coulomb criteria will be valid.

Fine grained saturated soils will respond to loading initially by generating excess pore water pressures and remaining at constant volume. At this stage the Tresca criteria, which uses total stress to represent undrained behaviour, should be used. This is the short term or immediate loading response. Once the pore pressure has dissapated, after a certain time, the effective stresses have incresed and the Mohr-Coulomb criterion will describe the strength mobilised. This is the long term loading response.

Tresca Criterion

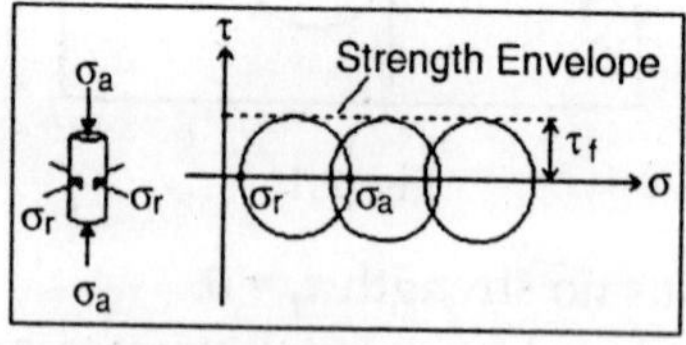

Fig. 2.14

The strength is independent of the normal stress since the response to loading simple increases the pore water pressure and not the effective stress.

The shear strength τ_f is a material parameter which is known as the undrained shear strength σ_u.

$$\tau_f = (\sigma_a - \sigma_r) = \text{constant}$$

Mohr-Coulomb (c'=0) criterion

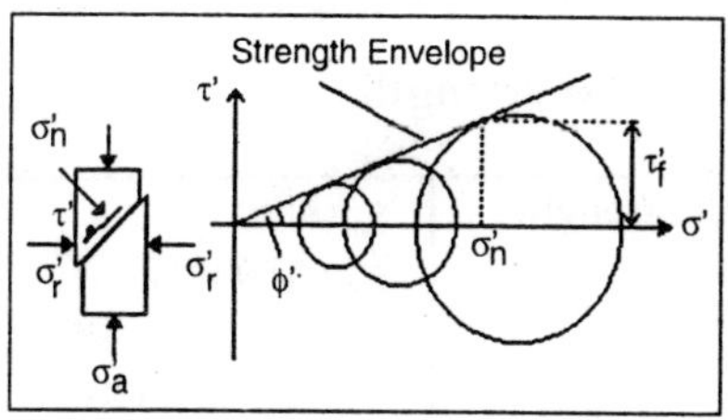

Fig. 2.15

The strength increases linearly with increasing normal stress and is zero when the normal stress is zero.

$$\tau'_f = \sigma'_n \tan\phi'$$

ϕ' is the angle of friction

In the Mohr-Coulomb criterion the material parameter is the angle of friction f and materials which meet this criterion are known as frictional. In soils, the Mohr-Coulomb criterion applies when the normal stress is an effective normal stress.

MOHR-COULOMB (C'>0) CRITERION

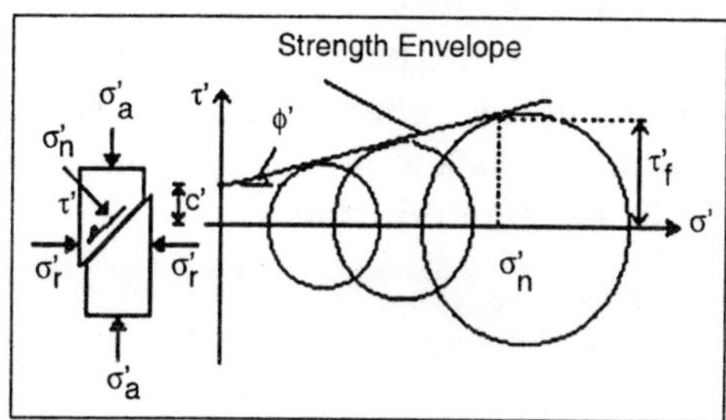

Fig. 2.16

The strength increases linearly with increasing normal stress and is positive when the normal stress is zero:

- $\tau'_f = c' + \sigma'_n \tan\phi'$
- ϕ' is the angle of friction
- c' is the 'cohesion' intercept

In soils, the Mohr-Coulomb criterion applies when the normal stress is an effective normal stress. In soils, the cohesion in the effective stress Mohr-Coulomb criterion is not the same as the cohesion (or undrained strength s_u) in the Tresca criterion.

Typical values of Shear strength

Undrained shear strength	s_u (kPa)	
Hard soil	s_u > 150 kPa	
Stiff soil	s_u = 75 ~ 150 kPa	
Firm soil	s_u = 40 ~ 75 kPa	
Soft soil	s_u = 20 ~ 40kPa	
Very soft soil	`s_u < 20 kPa	
Drained shear strength	c′ (kPa)	ϕ′ (deg)
Compact sands	0	35°–45°
Loose sands	0	30°–35°
Unweathered overconsolidated clay		
critical state	0	18° ~ 25°
peak state	10 ~ 25 kPa	20° ~ 28°
residual	0 ~ 5 kPa	8° ~ 15°

Often the value of c′ deduced from laboratory test results (in the shear testing apperatus) may appear to indicate some shar strength at σ′ = 0. *i.e.* the particles 'cohereing' together or are 'cemented' in some way. Often this is due to fitting a c′, ϕ′ line to the experimental data and an 'apparent' cohesion may be deduced due to suction or dilatancy.

Chapter 3

Soils, Weathering, and Nutrients

SOILS

All our elements, with the exception of hydrogen and helium, came from the dying throes of large stars. Amino acids needed to build animal proteins come from autotrophic plant life.

Biological studies have shown how certain nutrients (*e.g.*, nitrogen, phosphorus, calcium, magnesium, potassium, iron, etc.) are essential for the process of protein formation. All our amino acids and nutrients eventually come to us from plant life (sometimes via the meat of plant-eating animals). Plants synthesize amino acids from the combination of *sunlight, water and soils.*

Soil is therefore of critical importance to life. Simply put: no soil, no life. We first define soil as a *dynamic natural body capable of supporting a vegetative cover.* Where there is no soil, there is no plant life and we have barren rock and/or sand. Soil is composed primarily of *weathered* materials, along with water, oxygen and organic materials. Luckily for us, soil covers most of the land surface with a fragile, thin mantle. Soil and agricultural scientists have identified a huge number of different soil types.

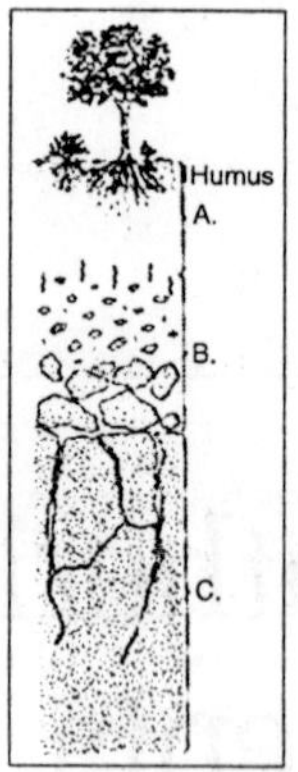

Fig. 3.1 Layers of Horizons of a Typical Soil Profile

SOIL IS LAYERED

Soil is layered into sections called "horizons". Figure 3.1 shows a typical soil profile developed on granite bedrock in a temperate region. The top horizon is composed of *humus* and contains most of the organic matter. This layer is often the darkest. The "A" horizon consists of tiny particles of decayed leaves, twigs and animal remains. The minerals in the A-horizon are mostly clays and other insoluble minerals. Minerals that dissolve in water are found at greater depths.

The "B" horizon has relatively little organic material, but contains the soluble materials that are *leached* downwards from above. The "C" horizon is slightly broken-up bedrock, typically found 1-10 meters below the surface. While this is a typical soil profile, many other types exist, depending on climate, local rock conditions and the community of organisms living nearby. The U.S. Department of Agriculture has classified 10 orders and 47 suborders of soils. If you include other subsets, there are over 60,000 types of soil. The lunar surface, which has been produced by meteoroid impacts, is not classified as a soil, but is rather given the name "regolith" (derived from the Greek words meaning cover and stone). The layered nature of soil indicates its long evolution under the effects of atmospheric and biological processes. The process that creates soil from bare rock is called "weathering". In the weathering process, the atmosphere and water interact with bare rock to slowly break it down into smaller and smaller particles. Rock climbers who encounter talus slopes (regions of pebble-life rocks that form in great conical piles at the feet of mountains) experience an intermediate step in the inexorable transition from solid granite to sand and soil. We next discuss the process whereby bare volcanic rock can be slowly turned into soils that can support life.

SOIL DEVELOPMENT

Soil forms from a complex interaction between earth materials, climate, and organisms acting over time. The brightly coloured soils of the humid tropics reflect the intense chemical reactions occurring in warm climates. The fertile prairie soils of the American Midwest evolved from the nutrient-rich organic matter left by decaying grasses. Regardless of soil characteristics, the whole process starts with the breakdown of earth material.

Fig. 3.2

WEATHERING

Weathering refers to processes that physically breakdown and chemically alter earth material. *Physical weathering,* also known as *mechanical weathering,* is the breakdown of large pieces of earth material into smaller ones. Think of physical weathering as the *disintegration* of rock without changing its chemical composition. There are many ways earth material can be physically weathered. When water freezes in rock crevices it expands creating stress in the crevice. As the stress increases, the crevice widens ultimately breaking the rock. Plant roots wedge rocks apart as they grow into rock crevices too. The shrinking and swelling by alternating heating and cooling weakens mineral bonds causing the rock to disintegrate. A very important result of physical weathering is its impact on the surface area of weathered material. When a block of earth material is broken into several smaller pieces, the amount of exposed surface increases. Examine the diagram below. A block with a width, depth, and height of 1 cm has a total surface area of 6 square centimeters. If we break the block in half in all directions it yields eight smaller pieces all with width, height, and depth of .5 cm. Breaking the block apart creates additional exposed surfaces such that the total surface area is now 12 square centimeters. Having more total exposed surface provides more area upon which chemical reactions can take place to further weather the material. The shape of the pieces also affects the the amount of exposed surface area. Plate-like pieces have more exposed surface area than do block-like pieces.

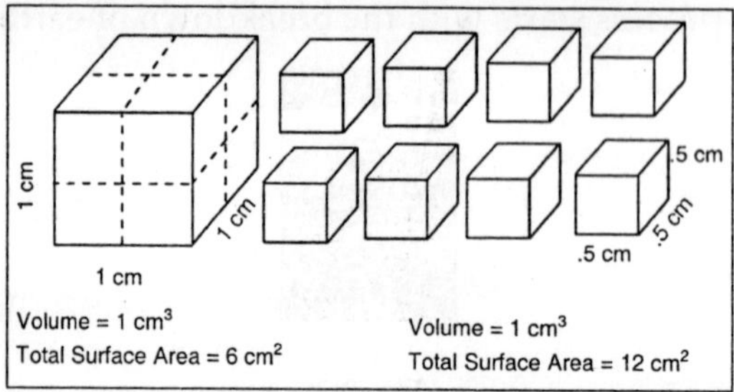

Fig. 3.3 Effect of Physical Weathering on Surface Area

Chemical weathering breaks down earth material by chemical alteration. This usually means adding a substance like water or air to the material. For instance, when oxygen is added to iron bearing minerals, *oxidation* takes places and a loose mantle of iron oxide is created (rust). *Hydrolysis* is an exchange reaction involving minerals and water. Free hydrogen (H^+) and hydroxide $(OH)^-$ ions in water replace mineral ions and drive them into solution. As a result, the mineral's structure is changed into a new form. Hydrolysis is a common process whereby silicate minerals are weathered into a clay mineral. Think of chemical weathering as the decomposition of earth material.

The spatial variation of climate and organisms play a significant role in the weathering of earth materials. Dry locations tend to be dominated by physical weathering and moist places by chemical weathering. The type of earth material available also determines the amount of weathering that might take place. Limestone is easily broken down where abundant rainfall and high temperatures prevail. However, limestone will remain intact in dry locations. The end result of the weathering process is the creation of a *weathered mantle*. The weathered mantle is not yet a soil until it undergoes further change. This involves the addition, transformation, translocation and removal of materials from the weathered mantle to form distinctive soil layers.

Since early geologic time, the atmosphere has interacted with the Earth's exposed crust though a process known as *weathering*. Weathering takes place through a combination of both mechanical and chemical means. We have all experienced the results of weathering first hand. Any visit to an old cemetery find us peering at the blurred inscriptions on old marble tombstones. These inscriptions were once perfectly legible, but with the passage of time, the small fractures and cracks in the rock have made it vulnerable to attack by aqueous solutions. A dramatic example of weathering can be seen in these two images of the same 3000 year old Egyptian obelisk just before relocation to damp New York and 100 years after accelerated weathering in New York Weathering rates are obviously a strong function of climate!

Fig. 3.4

Many of the original volcanic gases (*e.g.*, carbon dioxide, sulphur-bearing gases, etc.) were able to dissolve in water and produce acids. The acids, in turn, reacted with surface minerals. Later, oxygen in the atmosphere reacted with the exposed reduced materials, making the red beds discussed in an earlier lecture. Since the advent of land plants, soil and surface minerals have been exposed to relatively high concentrations of carbon dioxide maintained in soil pores as a result of decomposition and the metabolic activities of roots. The reaction of carbon dioxide with water in the soil produces carbonic acid (H_2CO_3) which determines the rate of rock weathering in most ecosystems.

$$CO_2 \text{ (gas)} + H_2O \text{ (liquid)} - H_2CO_3 \text{ (solution)}$$

Acid rain, produced by human effluents of nitrogen and sulphur-bearing gases will increase the rate of rock weathering in downwind areas. To understand why weathering occurs and why the rate of rock weathering is so dependent on climate, we need to discuss the chemistry of the process in a little more detail.

IGNEOUS ROCK WEATHERING

There is a well known expression that captures much of the story of rock weathering and illustrates the important role it has played in Earth history:

"Igneous Rocks + Acid Volatiles = Sedimentary Rocks + Salty Oceans"

What we mean by this will become clearer if we look at the details of the weathering processes. Consider a boulder or rock containing *Feldspar* minerals. Feldspar is a general term for a group of aluminosilicate minerals containing sodium, calcium, or potassium and having a lattice framework structure that makes for rigidity. Feldspars turn out to be one of the most common minerals in the Earth's crust. Feldspars are weathered through the chemical process of *hydration:*

$$K\,Al\,Si_3O_8 + H_2O \rightarrow Al_2SiO_5(OH)_4$$

In this chemical formula feldspar reacts with water to produce a kaolinite (clay). Notice that the chemical equation does not exactly balance, that is, not all the elements on the left hand side appear on the right hand side. This is because soluble elements, such as potassium (K) are *leached* out during the chemical reaction and carried away as dissolved salts. The process of leaching can perhaps best be understood by analogy with the making of coffee. When hot water is passed over crushed coffee beans, the soluble components (making the coffee) are leached away, leaving the insoluble crushed coffee bean remnants behind.

In this way, rocks containing feldspars are weakened though the conversion of rigid feldspar to more plastic clays which do not have anything like the same structural rigidity. The process occurs at exposed surfaces of the minerals making up the rock. Through geologic time, large amounts of sedimentary r0ocks have been deposited as part of this process. In fact about 75% of all exposed rocks on the Earth's surface today are of sedimentary origin and have been brought to the surface by geologic uplift.

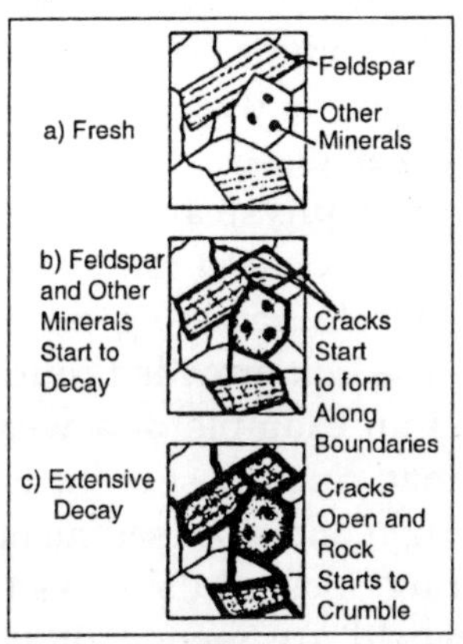

Fig. 3.5

Of course, geological processes return some of the sedimentary rocks to the mantle of the Earth, where they are converted back to the primary materials under conditions of great temperature and pressure. Rock weathering is also critical for the release of biochemical elements that have no gaseous form-examples are calcium, Ca, Potassium, K, Iron, Fe, and Phosphorus, P. The latter element plays a key role in cell metabolism. Thus, we can say that weathering provides key nutrients for life through the process of leaching.

The soluble nutrients are transferred to soils layers below the immediate surface (*e.g.*, the "B" layer). This is a major reason that plants have evolved root systems-to search out these critically needed nutrients below ground.

Igneous rock weathering proceeds in stages. It is useful to picture the inside of a rock-made up of interlocking minerals of irregular shapes, each being rigid. Figure 3.5 shows a microscopic view of such an interior rock composition, with some of the grains being feldspars.

As weathering proceeds, the boundaries of the vulnerable feldspar (and other) mineral grains start to decay. As the decay proceeds, water can reach more and more feldspar surfaces and the process accelerates. This process can also be accelerated by melt-freeze cycles that force the grains apart due to the difference between the volume occupied by water and ice. We can see that weathering is due to the combined effects of chemical and mechanical decay.

The chemical and physical processes of weathering transform the igneous rock into sand and clay particles and dissolved salts. Chemical weathering can add carbon dioxide, water, and oxygen. The link provided courtesy of the National Park Service shows an example of a weathering rock. It is interesting to note that, since most of the Earth's exposed rock is of sedimentary origin and since sedimentary rocks are a by-product of weathering, most of the rocks that are weathering away in today's world are second or perhaps even third generation rocks.

That is, they originated as igneous material, became sedimentary through weathering and transport to the bottom of shallow waters, were then subject to geologic uplift to become again exposed and, finally, began to undergo weathering yet again. The natural world is full of such endless cycles.

HOW FAST DOES A ROCK DECAY

The best answer to this question is... it depends. It depends on the local environment and the type of rock. For example, an iron nail buried in the ground in Michigan will only take a year or so to decay to the point that it is easily snapped in two. Iron nails rust much more slowly in drier environments. Aluminum cans decay very slowly, even in humid climates. Glass decays even more slowly, while plastic is considered essentially non-biodegradable. It is somewhat ironic that a plastic tombstone will endure much longer than one made in marble!

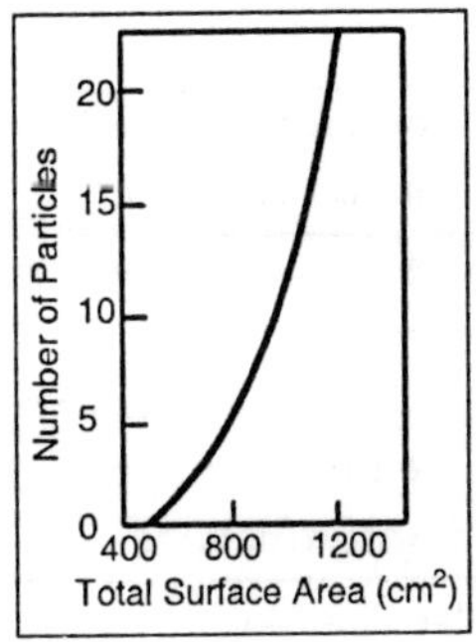

Fig. 3.6

From what we have already discussed, soils themselves aid in rock decay, as do melt-freeze cycles and bacterial action. Thus, soils are a consequence of weathering, but also a factor in accelerating weathering. The production of soil is a *positive feedback* process. The following table illustrates rates of weathering for three rock types as a function of climate. As more of a rock becomes amenable to weathering, the speed of weathering increases. This can be understood if we plot the rate at which the available exposed surface area of a rock increases as the rock fragments. Figure 3.6 shows this relationship.

NUTRIENTS

Although living tissue is composed of carbon, hydrogen, and oxygen in the approximate proportion of CH_2O, as many as 23 other elements are necessary for biochemical reactions and for the growth of structural biomass.

Examples of important nutrients are:

Table. Rates of Weathering of Clean Rock Surfaces (Micro-Meters/1000Years)

Rock Type	Cold Climate	Warm, Humid Climate
Basalt	10	100
Granite	1	10
Marble	20	200

Nutrient	Role
Nitrogen	The proteins found in plants and animals contain about 16% by weight of nitrogen
Phosphorus	Pis part of the important ribulose biphosphate carboxylase molecule and is part of ATP-adenosine triphosphate, the universal molecule for energy transformations
Calcium	Ca is a major structural component of the proteins forming plants and animals

Other important nutrients include magnesium, potassium, iron, sulphur, etc. We should note that, although carbon, nitrogen and sulphur can be obtained from the atmosphere, calcium, magnesium, potassium, iron, and phosphorus all come from rock weathering processes. The atmosphere has no store of these essential nutrients

As mentioned earlier, one of the main purposes of plant root systems is to get access to nutrients stored in the soil. Some plants go to enormous lengths to do this.

For example, Emiliani quotes that "A single plant of winter rye, 50 cm high, was found to have a root system consisting of 143 main roots, 35,600 secondary roots, 2.3 million tertiary roots, and 11.5 million quaternary roots! The root system was found to have a total length of 600 km and a total surface of about 250 square meters". Delivery of nutrients into plant root systems can occur by several pathways. In some cases, direct uptake in water solution occurs. Sometimes, plants actually have to protect themselves against too much nutrient intake.

Too much of a good thing can prove poisonous. An example of this can be seen in the accumulations of calcium carbonate deposits that surround the roots of some desert shrubs.

Table. Nature's Vitamin Requirements

Amount	Nutrient
100 parts	N
15 parts	P
50 parts	K
5 parts	Ca
5 parts	Mg
10 parts	S

Some nutrients, such as nitrogen, phosphorus, and potassium are often harder for roots to find and specialized (incredibly efficient) enzymes have evolved located in root membranes to seek out these scarce and needed resources. If some nutrients are not readily available, plants will grow more slowly and/or increase their root/shoot ratio.

In general, the availability of nutrients (deficit or surplus availabil ity) often controls the form of the ecosystem, determining its overall productivity, and influencing which particular set of plants come to predominate. Excess nitrogen can lead to the loss of fine root biomass and deficiencies in other nutrients.

The pool of nutrients held in the soil and vegetation is many times larger than the annual receipt of nutrients from the atmosphere and rock weathering. Thus, life *husbands* its needed nutrients on land, storing much of the total in the humus. Recycling of nutrients is critical to the productivity of natural ecosystems, although less critical to crop production, due to the availability of commercial fertilizers.

Table.Percentage of Annual Nutrient Requirement for Growth of Hardwood Forest

Process (kg/ha/yr)	N	P	K	Ca	Mg
Total Growth Requirement	115	12	67	62	10
Atmospheric Inputs	18	0	1	4	6
Rock Weathering Inputs	0	13	11	34	37
Reabsorptions (Intra-system)	31	28	4	0	2
Detritus Turnover	69	81	86	85	87

The data of the table came from a study of the famous Hubbard Brook ecosystem in New Hampshire. It shows how effective the soils are in storing the needed nutrients and how limited are the rates of supply from atmosphere and rock weathering. Acid rain caused by human emissions of nitrogen and sulphur oxides leads to enhanced weathering and changes in nutrient ratios. For example, recent studies suggest that forest growth has declined in areas downwind of air pollution. Acid rain appears to increase the movement of aluminum ions, which may, in turn, reduce the intake rates of calcium and other nutrients.

In the oceans, life is also limited by the availability of nutrients. The productivity is highest on continental shelves and in regions of upwelling. Nutrients are removed from surface waters by downward sinking and are regenerated in deep waters. A paucity of nutrients limits production in the open oceans. On the other hand, marine productivity is threatened by excessive human inputs of nitrogen and phosphorus in coastal regions, where it leads to excessive algal blooms, whose subsequent decay robs deeper water of oxygen, creating "dead zones"

SOIL EROSION

We define soil erosion as the movement of surface litter and topsoil. The forces responsible are wind and water flow. It is important to recognize at the outset that soil erosion is a natural process. It is slowed, however, by plant roots that serve to stabilize the soil. In any undisturbed ecosystem, the rate of soil loss is matched by its rate of production. In a disturbed ecosystem, however, major changes in the rate of erosion can occur. In fact, almost every activity that can be characterized as a "development" causes enhanced soil erosion. This includes farming, logging, building, grazing, off-road travel, etc. Soil erosion, discussed later in this lecture, is responsible for the loss of a great deal of our topsoil.

The problem of accelerated soil erosion is a major one. On a global scale, topsoil is eroding faster than it can be replenished in over one third of the world's croplands. For example, in China and India combined, more than 12 million square kilometers have been severely eroded since 1945. The causes of this have been deforestation (30%), overgrazing (35%), and farming (28%). The soil erosion problem is particularly severe if one considers that each year we must feed an extra 90 million more people, with about 25 billion tons less topsoil!

In the U.S., the situation is also of concern. The Dust Bowl of the 1930's was caused by plowing of the prairies. Before the pioneers, the soil was held in place by the long root systems of prairie grasses. In response to the dust bowl, the Soil Conservation Service (SCS) was established in 1935. The Dust Bowl was brought on by a long drought and lasted about a decade. Much was learned about soil conservation and later droughts have not caused as much damage.

Fig. 3.7

The Great Plains has lost about one-third of its original topsoil in the past 150 years. Iowa has been particularly hard hit, already having lost half of its topsoil since the arrival of the first European settlers. In California, the current topsoil is being eroded at a rate that exceeds its replenishment rate by a factor of more than 70. The amount of topsoil lost in a day is staggering: Emiliani estimates that the lost topsoil would fill a line of dump trucks 3,500 miles long! The cost of this loss of natural resource is incalculable. The map below summarizes points brought up in lecture on geographic distribution of soil degradation:

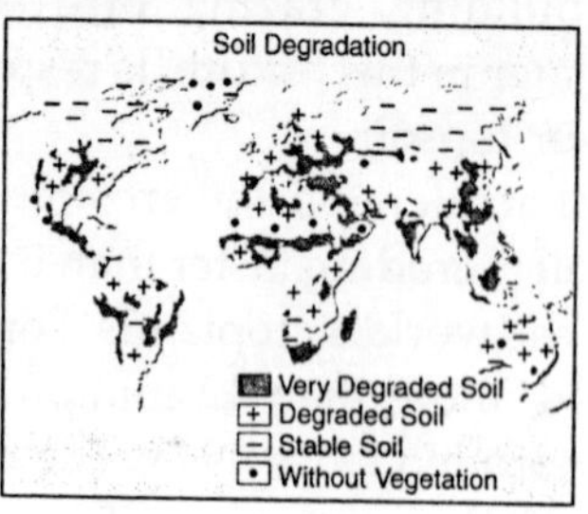

Fig. 3.8

Soil is important for the support of life processes. Real soil only exists on Earth. Soil is the place where atmosphere, biosphere, hydrosphere, and lithosphere meet. Over 60,000 types of soils have been catalogued

Without soil, there is no food. Soils take 10's of thousands of years to form in rock weathering processes. Weathering occurs due to a combination of chemical and mechanical processes that are subject to strong positive (reinforcing) feedback mechanisms. Over geologic time, rock weathering processes have led to sedimentary rocks and salty oceans!

Plants obtain inorganic minerals (nutrients) from the soil and incorporate their elements into biochemical materials. Animals may eat plants and each other, and synthesize new proteins, but the building blocks are the amino acids originally synthesized in plants. Soil effectively husbands the needed nutrients in the upper horizons.

The availability (or lack of availability) of important nutrients places critical constrains on the type of life that can survive in particular ecosystems. Soil erosion is a very serious problem globally because of its inextricable link to development activities. Soil conservation is therefore a very serious business and an important part of future global change studies.

Chapter 4

Solid Phase Peptide Synthesis

INTRODUCTION

GENERAL PRINCIPLES

The revolutionary principle behind solid phase peptide synthesis (SPPS) as conceived by Merrifield in 1959 is that if the peptide is bound to an insoluble support then any unreacted reagents left at the end of any synthetic step can be removed by a simple wash procedure, greatly decreasing the time required for synthesis. What is more, the arrangement is amenable to automation. This is only valid, however, if the individual synthetic steps occur with essentially quantitative yields. This latter requirement entails rigorous testing of the chemistry and tends to result in the use of a limited range of tried and tested methodologies.

The two most commonly used schemes are the original Merrifield method and that proposed by Sheppard in 1975. While the protection schemes used are different, they have the same basic concepts, a resin support, the use of an excess of reagents and building the peptide in a C→N terminal direction.

The resin supports used are designed to swell in commonly used solvents, expanding to many times their original size. Thus, the reactions occur not on the surface of a rigid particle, but within the support in a solvated gel which permits easy access to the growing peptide chain.

Reagents are used in excess, this allows the reactions to proceed to completion in the minimum time and results in faster synthesis of peptides with high purity. The couplings proceed in a C→N terminal direction to allow the use of racemisation limiting amine protection for the activated species.

MERRIFIELD SYNTHESIS

This methodology is characterised by the use of *tert*-butyl based temporary a-amino protection and benzyl, or substituted benzyl, groups for permanent side chain protection. There are over one hundred different substituted resins suitable for peptide synthesis generally based on polystyrene and polyethylene glycol. These resins allow introduction of an amino acid through either substitution, condensation or addition reactions. The traditional resin used for Merrifield synthesis was a chloromethylphenyl substituted resin. The first amino acid was attached to the resin through substitution of the chloride by the caesium salt of the BOC-amino acid, generating an equivalent to a benzyl ester.

Deprotection of the temporary BOC group uses a 20-50% solution of trifluoroacetic acid (TFA) in dichloromethane and has to be followed by neutralisation of the resulting ammonium salt with a hindered tertiary base. Final cleavage from the resin as well as deprotection of benzyl based side chain protecting groups is achieved using strong acids, usually liquid hydrogen fluoride or trifluoromethane sulphonic acid. Such procedures require specialised apparatus and the highly acidic conditions catalyse several possible rearrangements.

FMOC POLYAMIDE SYNTHESIS

The fundamental differences between the Fmoc polyamide strategy when compared to the Merrifield approach are that the reactions are carried out under continuous flow and that the conditions for a-amino deprotection and cleavage from the resin are far more mild. This arises from the adoption of the base labile

Fmoc protecting group for a-amino protection. The side chains are generally protected with *tert*-butyl based groups which, in common with the linkage to the resin, can be cleaved by TFA in the presence of scavengers.

Traditionally, resins with 4 hydroxymethylphenoxy substitution were used. These allowed were esterified with the anhydride of the first amino acid. As a result of the mesomerically electron donating *para* oxygen atom stabilising the resultant carbocation, cleavage of the peptide from the resin occurs under more mild acid conditions, typically using trifluoroacetic acid with scavengers.

The use of continuous flow means that the reagents are passed through a reaction chamber containing the resin supported peptide. Having passed through this chamber, the reagents can be recirculated back into the chamber again or taken to a waste collection bottle. This allows the resin to be washed clean of excess reagents and unwanted reaction products, which in turn helps drive deprotection steps to completion following Le Chatelier's principle. In addition, the solution can be passed through a u.v. detector and monitored at a suitable wavelength for the Fmoc chromophore.

A typical cycle consists of:

- Deprotection of the preceding residue with piperidine.
- Wash to remove any remaining reagents from 1.
- Acylation in a recirculatory mode.
- Wash to remove excess reagents.

A typical uv trace indicates these steps.

In this way, the cycle can be qualitatively monitored. Such monitoring can be used in conjunction with automation. For example, a slow deprotection step suggests inaccessibility of the peptide amino terminal, indicating that the next coupling step may require a longer acylation time. If this is detected then the cycle can be interrupted for later manual intervention.

Since the turn of the century, there has been a constant drive to develop new and improved strategies for peptide synthesis. The areas of protection, deprotection, activation and coupling have all received attention. The field of solid phase peptide synthesis has rapidly grown since its inception by Merrifield but is dominated by use of BOC/benzyl and Fmoc/*tert*-butyl protection schemes.

SOLUTION PHASE PEPTIDE SYNTHESIS

Care has to be taken in selection of protecting groups so that they can be removed with appropriate selectivity. Side chain and carboxyl protecting groups need to be selected so as not to be labile under the conditions required to deprotect the amino group. If a long peptide is to be made then it is common to synthesise it in smaller sections, which are later coupled to form the overall peptide-a process called fragment condensation. If a fragment condensation technique is being adopted then the carboxyl terming must also be deprotected without loss of side chain protection.

There are two ways of achieving this selective deprotection:

1. Chose protecting groups that are deprotected with completely different reagents (refered to as orthogonal protection)
   e.g. tert butyl (acid), fluorenyl methyl (base), benzyl (catalytic hydrogenolysis).
2. Chose protecting groups that are deprotected with the same type of reagent but under different conditions. *e.g. tert* butyl and benzyl which require increasingly strong acids for deprotection.

SOIL ORGANIC MATTER

On the basis of organic matter content, soils are characterized as mineral or organic. Mineral soils form most of the world's cultivated land and may contain from a trace to 30 per cent organic matter. Organic soils are naturally rich in organic matter principally for climatic reasons. Although they contain more than 30 per cent organic matter, it is precisely for this reason that they are not vital cropping soils. This soils bulletin concentrates on the organic matter dynamics of cropping soils. In brief, it discusses circumstances that deplete organic matter and the negative outcomes of this. The bulletin then moves on to more proactive solutions. It reviews a "basket" of practices in order to show how they can increase organic matter content and discusses the land and cropping benefits that then accrue. Soil organic matter is any material produced originally by living organisms (plant or animal) that is returned to the soil and goes through the decomposition process. At any given time, it consists of a range of materials from the intact original tissues of plants and animals to the substantially decomposed mixture of materials known as humus.

Fig. 4.1 Plate Crop Residues Added to The Soil are Decomposed by Soil Macrofauna and Micro-Organisms, Increasing The Organic Matter Content of The Soil.

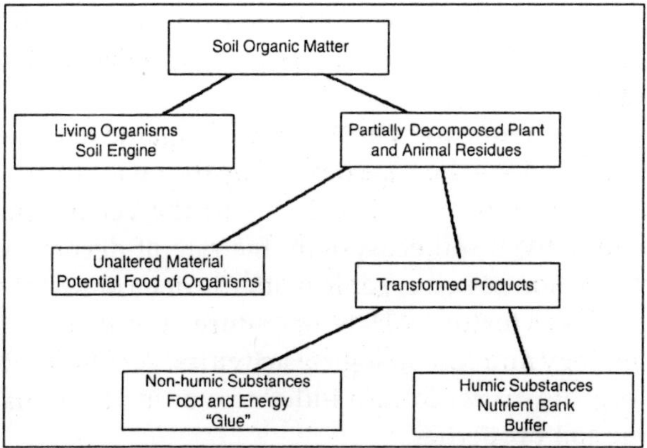

Fig. 4.2 Components of Soil Organic Matter and Their Functions

Most soil organic matter originates from plant tissue. Plant residues contain 60-90 per cent moisture. The remaining dry matter consists of carbon (C), oxygen, hydrogen (H) and small amounts of sulphur (S), nitrogen (N), phosphorus (P), potassium

(K), calcium (Ca) and magnesium (Mg). Although present in small amounts, these nutrients are very important from the viewpoint of soil fertility management.

Soil organic matter consists of a variety of components. These include, in varying proportions and many intermediate stages, an active organic fraction including microorganisms (10-40 per cent), and resistant or stable organic matter (40-60 per cent), also referred to as humus.

Forms and classification of soil organic matter have been described by Tate and Theng. For practical purposes, organic matter may be divided into aboveground and belowground fractions. Aboveground organic matter comprises plant residues and animal residues; belowground organic matter consists of living soil fauna and microflora, partially decomposed plant and animal residues, and humic substances. The C:N ratio is also used to indicate the type of material and ease of decomposition; hard woody materials with a high C:N ratio being more resilient than soft leafy materials with a low C:N ratio.

Although soil organic matter can be partitioned conveniently into different fractions, these do not represent static end products. Instead, the amounts present reflect a dynamic equilibrium.

The total amount and partitioning of organic matter in the soil is influenced by soil properties and by the quantity of annual inputs of plant and animal residues to the ecosystem. For example, in a given soil ecosystem, the rate of decomposition and accumulation of soil organic matter is determined by such soil properties as texture, pH, temperature, moisture, aeration, clay mineralogy and soil biological activities. A complication is that soil organic matter in turn influences or modifies many of these same soil properties.

Organic matter existing on the soil surface as raw plant residues helps protect the soil from the effect of rainfall, wind and sun. Removal, incorporation or burning of residues exposes the soil to negative climatic impacts, and removal or burning deprives the soil organisms of their primary energy source. Organic matter within the soil serves several functions.

From a practical agricultural standpoint, it is important for two main reasons:

1 As a "revolving nutrient fund"; and
2. As an agent to improve soil structure, maintain tilth and minimize erosion.

As a revolving nutrient fund, organic matter serves two main functions:

1. As soil organic matter is derived mainly from plant residues, it contains all of the essential plant nutrients. Therefore, accumulated organic matter is a storehouse of plant nutrients.
2. The stable organic fraction (humus) adsorbs and holds nutrients in a plant-available form.

Organic matter releases nutrients in a plant-available form upon decomposition. In order to maintain this nutrient cycling system, the rate of organic matter addition from crop residues, manure and any other sources must equal the rate of decomposition, and take into account the rate of uptake by plants and losses by leaching and erosion.

Where the rate of addition is less than the rate of decomposition, soil organic matter declines. Conversely, where the rate of addition is higher than the rate of decomposition, soil organic matter increases. The term steady state describes a condition where the rate of addition is equal to the rate of decomposition.

In terms of improving soil structure, the active and some of the resistant soil organic components, together with micro-organisms (especially fungi), are involved in binding soil particles into larger aggregates. Aggregation is important for good soil structure, aeration, water infiltration and resistance to erosion and crusting.

Traditionally, soil aggregation has been linked with either total C or organic C levels. More recently, techniques have developed to fractionate C on the basis of lability (ease of oxidation), recognizing that these subpools of C may have greater effect on soil physical stability and be more sensitive indicators than total C values of carbon dynamics in agricultural systems. The labile carbon fraction has been shown to be an indicator of key soil chemical and physical properties. For example, this fraction has been shown to be the primary factor controlling aggregate breakdown in Ferrosols (non-cracking red clays), measured by the percentage of aggregates measuring less than 0.125 mm in the surface crust after simulated rain in the laboratory.

The resistant or stable fraction of soil organic matter contributes mainly to nutrient holding capacity (cation exchange capacity [CEC]) and soil colour. This fraction of organic matter decomposes very slowly. Therefore, it has less influence on soil fertility than the active organic fraction.

WHAT IS ORGANIC MATTER

Soil organic matter consists of a variety of components. *These include, in varying proportions and many Intermediate stages:*

- Raw plant residues and microorganisms (1 to 10 per cent)
- "Active" organic traction (10 to 40 per cent)
- Resistant or stable organic matter (40 to 60 per cent) also referred to as humus.

Raw plant residues, on the surface, help reduce surface wind speed and water run-off. Removal, incorporation or burning of residues predisposes the soil to serious erosion. The "active" and some of the resistant soil organic components, together with microorganisms (especially fungi) are involved in binding small soil particles into larger aggregates. Aggregation is important for good soil structure, aeration, water infiltration and resistance to erosion and crusting. The resistant or stable fraction of soil organic matter contributes mainly to nutrient holding capacity (cation exchange capacity) and soil colour.

This fraction of organic matter decomposes very slowly and therefore has less influence on soil fertility than the "active" organic fraction. Organic matter in soil serves several functions. From a practical agricultural standpoint, it is important for two main reasons. First as a "revolving nutrient bank account"; and second, as an agent to improve soil structure, maintain tilth, and minimize erosion.

As a revolving nutrient bank account, organic matter serves two main functions:

1. Since soil organic matter is derived mainly from plant residues, it contains all of the essential plant nutrients. Accumulated organic matter, therefore, is a storehouse of plant nutrients. Upon de-composition, the nutrients are released in a plant-available form.
2. The stable organic fraction (humus) adsorbs and holds nutrients in a plant available form.

Organic matter does not add any "new' plant nutrients but releases nutrients in a plant available form through the process of decomposition. In order to maintain this nutrient cycling system, the rate of addition from crop residues and manure must equal the rate of decomposition. If the rate of addition is less than the rate of decomposition, soil organic matter will decline and, conversely if the rate of addition is greater than the rate of decomposition, soil organic matter will increase.

The term steady state has been used to describe a condition where the rate of addition is equal to the rate of decomposition. Fertilizer can contribute to the maintenance of this revolving nutrient bank account by increasing crop yields and consequently the amount of residues returned to the soil.

ORGANIC MATTER IN VIRGIN AND CULTIVATED SOILS

Soils in Alberta are divided into soil groups (zones) based on the amount of organic matter they contain. They occur in geographic zones from the southeast to the northwest and are identified as the Brown, Dark Brown, and Black Chernozemic (prairie) soils. The Brown soils have the least amount of organic matter because of the relatively small inputs of plant residues contributed by the short grass prairie vegetation under which these soils developed. Black soils developed under cooler and wetter conditions which allowed for more grass growth and thus a greater accumulation of organic matter. Further north and west, trees became the dominant vegetation. Soils influenced by forest vegetation for a moderate length of time constitute the Dark Gray or transitional soils. Where the forest cover was established for a longer period, Luvisolic (forest) soils developed. Organic (peat) soils. occur in low lying areas throughout the Black, Dark Gray and Gray soil zones. These soils are saturated with water for much or all of the year thereby reducing the rate of organic matter decomposition. The amount of soil organic matter characteristic of virgin and cultivated soils in the various zones is shown in Table 1. Cultivation generally has resulted in a 30 to 50 per cent loss of organic matter.

Table. Organic Matter in Native and Cultivated Soils (per cent)

Soil zone	Virgin	Cultivated
Brown	3-4	2-3
Dark Brown	4-5	3-4
Black	6-10	4-6
Dark Gray	4-5	2-3
Gray	1-2	1-2

Before our soils were cultivated, they had achieved a "steady state". In most of our prairie soils, the increased rate of decomposition associated with cultivation, combined with the low rates of crop residue addition associated with crop-fallow rotations has caused a fairly rapid decline in soil organic matter. The rate of decline decreases with time as the amount of total soil organic matter decreases and particularly as the "active" organic fraction is depleted. Cultivation of soils that are naturally high in organic matter will usually result in a decrease of organic matter. In the case of Luvisolic soils, their poor physical properties and low fertility have encouraged the use of forages, fertilizers, manure and judicious tillage. Such management practices have resulted in an increase in soil organic matter on Luvisolic soils, whereas excessive tillage, fallowing and minimal fertilization have lead to further depletion of the soil organic matter.

EFFECTS OF ORGANIC MATTER DECLINE

As stated in the introduction, soil degradation is becoming a major concern in Canada. Loss of organic matter is often identified as one of the main factors contributing to declining soil productivity, but it is misleading to equate a loss in soil organic matter with a loss in soil productivity. Soil organic matter contributes to soil productivity in several ways, but there is no direct quantitative relationship between soil productivity and total soil organic matter. In fact, it has been the decline in organic matter that has contributed to the productivity of the crop-fallow system.

This decline in organic matter has resulted in the release of large amounts of plant nutrients, particularly nitrogen. For example, a decrease in soil organic matter of 2 per cent releases about 2,400 lb/ac of nitrogen. If this decline occurred over a 60 year period, an average of 40 lb/ac/yr of plant-available nitrogen has come from the soil organic matter. We therefore view prairie soils which had relatively high levels of organic matter as being nitrogen fertile, but this fertility could only be attained under a management system that allowed for organic matter to decline.

Frequent fallowing has been a major factor contributing to this decline. Insofar as organic matter contributes to improved soil physical properties (*e.g.*, tilth, aggregation, moisture holding capacity and resistance to erosion) increasing soil organic matter will generally result in increased soil productivity. But on many soils, suitable soil physical properties occur at relatively low levels of organic matter (2-4 per cent). A level of organic matter higher than required to produce suitable physical properties is beneficial in that the soil has a greater buffering and nutrient holding capacity, but it does not contribute directly to soil productivity. If soils are managed so organic matter is not declining (steady-state), soils higher in organic matter (*e.g.*, 8 per cent) are not inherently more productive or fertile than those that have less organic matter (*e.g.*, 5 per cent). To equate the ability to supply nutrients with total soil organic matter is not valid. The "active" fraction of organic matter is a more reliable indicator of soil fertility than is total soil organic matter. In cultivated soil, the "active" fraction is influence mainly by previous management. Soil organic matter cannot be increased quickly even when management practices that conserve soil organic matter are adopted. The increased addition of organic matter associated with continuous cropping, and the production of higher crop yields, are accompanied by an increase in the rate of decomposition. Moreover, only a small fraction of crop residues added to soil remains as soil organic matter. After an extended period of time, the return of all crop residues and the use of forages in rotations with cereals and oilseeds may significantly increase soil organic matter, particularly, the "active" fraction.

MANAGING SOIL ORGANIC MATTER

There have been vast changes in the nature of agricultural production. In the past, farms were small, and much of what was produced was consumed on the farm. This system allowed for the limited removal of soil nutrients since there was an opportunity to return most of the nutrients back to the land. The advent of the internal combustion engine, migration from rural to urban communities, increasing farm size and specialization in production have resulted in a system of production where there is greater removal of plant nutrients from the soil and less opportunity for nutrient cycling. Maintenance of organic matter for the sake of maintenance alone is not a practical approach to farming. It is more realistic to use a management system that will give sustained profitable production. The greatest source of soil organic matter is the residue contributed by current crops. Consequently, crop yield and type, method of handling residues and frequency of fallow are all important factors. Ultimately, soil organic matter must be maintained at a level necessary to maintain soil tilth.

Summerfallow

Summerfallowing accelerates the loss of organic matter. Aeration of the soil associated with tillage, and the increase in soil temperature and moisture results in increased organic matter decomposition. Since little In the way of residues are added to the soil, a net loss of organic matter occurs. Research has shown that as the frequency of fallow increases, the amount of soil organic matter decreases.

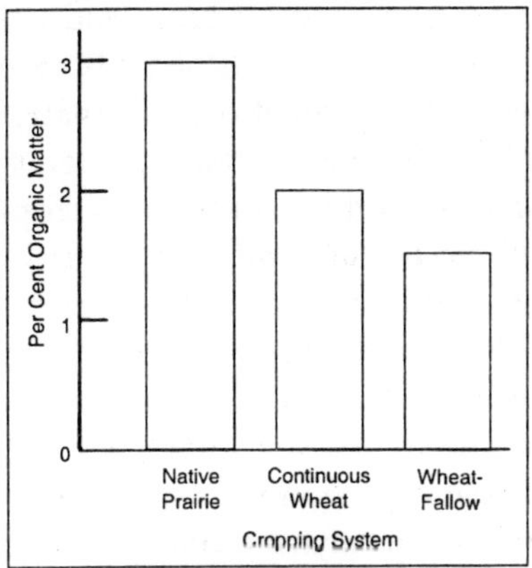

Fig. 4.3 Effect of Frequency of Fallow on Per Cent Organic Matter.

Summer fallowing for moisture conservation may be a necessary practice in the Brown and Dark Brown soil zones. However, it must be questioned in the Black and the Gray soil zones. Periodic fallowing may be acceptable in the higher rainfall regions for control of persistent perennial weeds and volunteer grains in pedigreed seed production. In the crop-fallow system common to the prairie region, the nitrogen removed has far exceeded that gained from crop residues, manure, legumes and fertilizer. The large reserves of nitrogen present in the organic matter of our prairie soils have been the major source of nitrogen in this cropping system. Continued reliance on soil organic matter reserves to supply the nitrogen requirements of crops will ultimately lead to a decline in soil productivity, and increased soil erosion. When a change from a cropping system involving fallow to continuous cereal grain production, the nitrogen requirement increases. The nitrogen requirement is greatest in the first few years of continuous cropping as the nutrient cycling process adjusts to the new cropping system.

Crop Rotations

The value of forage crops in rotations with cereals and

oilseeds has long been recognized, especially in the Luvisolic soils. The Breton Plots compared a two-year fallow-wheat rotation with a five-year rotation involving wheat, oats and barley followed by two years of hay production. In research done by the University of Manitoba, the effects of various cultural practices on the level of organic matter are compared. It is interesting to note that the highest level of soil organic matter was maintained under continuous cropping.

The beneficial effects of perennial forages are the result of:

- A more extensive root system and crop aftermath contributing more organic matter to the soil,
- The fibrous nature of the root system of perennial grasses. These are particularly effective as a binding agent in soil aggregation,
- Nitrogen fertility enhancement by the growth of legumes,
- Increased permeability of dense subsoils because of the deep penetrating tap roots of perennial legumes, especially alfalfa.
- A reduced rate of organic matter decomposition in the absence of tillage.

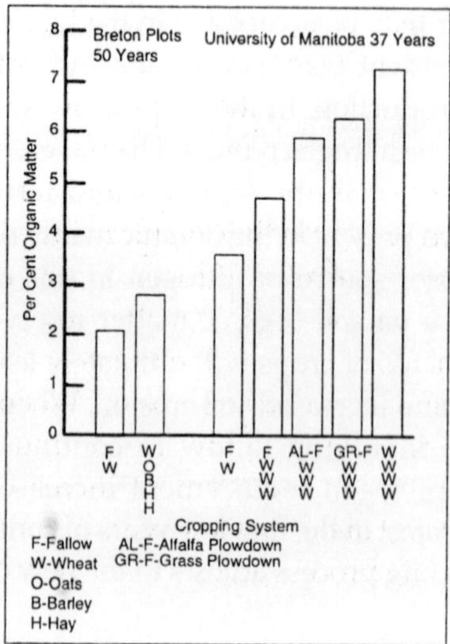

Fig. 4.4 Effect of Rotation on Soil Organic Matter

In the brown and dark brown soil zones, perennial forages grown for forage production or as a plowdown crop may jeopardize subsequent cereal crops because they deplete soil moisture reserves.

Fertilization

Fertilizers will generally increase soil organic matter because the increased crop growth returns larger amounts of residues to the soil. Data obtained from the Breton Plots is summarized in Figure 4.4. The increase in organic matter is less than what might be expected with current farming practices since all the straw had been removed from the plots.

To determine the fertilizer effect, two fertilizer treatments were averaged, one involving a low rate of nitrogen and sulphur and another involving a low rate of nitrogen, phosphorus, potassium and sulphur. One would expect that with higher rates of fertilization, higher yielding varieties, and the return of all crop residues, the effect of fertilizers on organic matter would be greater than that shown in Figure 4.5.

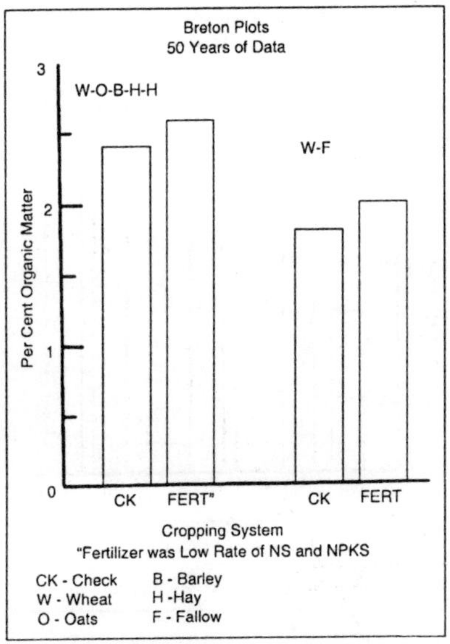

Fig. 4.5 Effect of Fertilizer on Soil Organic Matter.

Plowdown

Legume plowdown has received considerable attention in recent years as an alternative to the use of nitrogen fertilizers. However, when considering this option in a cropping programme, the amount of nitrogen added by the legume, as well as the loss of one year of production, the cost of seed and the expected yield increase must be kept in mind. Strictly as a source of nitrogen, the value of a legume plowdown is questionable.

The amount of nitrogen fixed by a legume is dependent upon the type of legume, the amount of vegetative growth, the nature of the soil and environmental conditions. As a source of organic matter, legume plowdown is valuable, however, perennial forage is more effective than legume plowdown for increasing soil organic matter. Nitrate nitrogen which accumulates following legume plowdown is subject to loss, particularly in wet, poorly drained soils. To minimize this, legumes should be plowed down in the fall rather than mid-summer to reduce nitrate accumulation and subsequent loss.

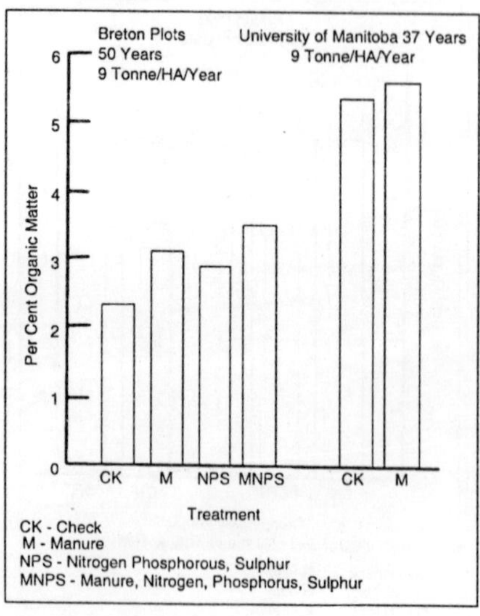

Fig. 4.6 Effect of Manure on Soild Organic Matter.

The cultivation of prairie soils has generally resulted in a decline in organic matter of 30 to 50 per cent. A product of this decline has been the release of large amounts of plant nutrients, particularly nitrogen. Crop rotations with a high frequency of summerfallow have relied on the nitrogen released from soil organic matter to supply crop requirements. More frequent or continuous cropping, less frequent tillage, the production of high yields and the return of crop residues will help to maintain soil organic matter at a satisfactory level. Perennial forages are effective for maintaining or increasing soil organic matter.

ORGANIC MATTER DECOMPOSITION AND THE SOIL FOOD WEB

When plant residues are returned to the soil, various organic compounds undergo decomposition. Decomposition is a biological process that includes the physical breakdown and biochemical transformation of complex organic molecules of dead material into simpler organic and inorganic molecules.

The continual addition of decaying plant residues to the soil surface contributes to the biological activity and the carbon cycling process in the soil. Breakdown of soil organic matter and root growth and decay also contribute to these processes. Carbon cycling is the continuous transformation of organic and inorganic carbon compounds by plants and micro- and macro-organisms between the soil, plants and the atmosphere.

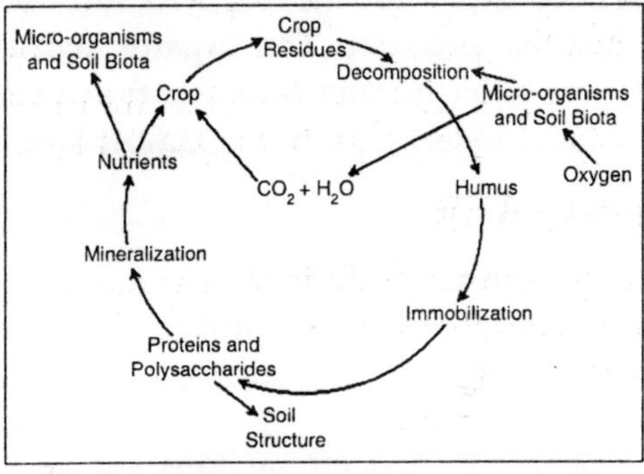

Fig. 4.7 Carbon Cycle

Decomposition of organic matter is largely a biological process that occurs naturally. Its speed is determined by three major factors: soil organisms, the physical environment and the quality of the organic matter. In the decomposition process, different products are released: carbon dioxide (CO_2), energy, water, plant nutrients and re-synthesized organic carbon compounds. Successive decomposition of dead material and modified organic matter results in the formation of a more complex organic matter called humus.

This process is called humification. Humus affects soil properties. As it slowly decomposes, it colours the soil darker; increases soil aggregation and aggregate stability; increases the CEC (the ability to attract and retain nutrients); and contributes N, P and other nutrients.

Soil organisms, including micro-organisms, use soil organic matter as food. As they break down the organic matter, any excess nutrients (N, P and S) are released into the soil in forms that plants can use. This release process is called mineralization. The waste products produced by micro-organisms are also soil organic matter. This waste material is less decomposable than the original plant and animal material, but it can be used by a large number of organisms.

By breaking down carbon structures and rebuilding new ones or storing the C into their own biomass, soil biota plays the most important role in nutrient cycling processes and, thus, in the ability of a soil to provide the crop with sufficient nutrients to harvest a healthy product. The organic matter content, especially the more stable humus, increases the capacity to store water and store (sequester) C from the atmosphere.

THE SOIL FOOD WEB

The soil ecosystem can be defined as an interdependent life-support system composed of air, water, minerals, organic matter, and macro- and micro-organisms, all of which function together and interact closely.

The organisms and their interactions enhance many soil ecosystem functions and make up the soil food web. The energy needed for all food webs is generated by primary producers: the plants, lichens, moss, photosynthetic bacteria and algae that use sunlight to transform CO_2 from the atmosphere into carbohydrates. Most other organisms depend on the primary producers for their energy and nutrients; they are called consumers.

Some functions of a healthy soil ecosystem:

- Decompose organic matter towards humus.
- Retain N and other nutrients.
- Glue soil particles together for best structure.
- Protect roots from diseases and parasites.
- Make retained nutrients available to the plant.
- Produce hormones that help plants grow.
- Retain water.

Soil life plays a major role in many natural processes that determine nutrient and water availability for agricultural productivity. The primary activities of all living organisms are growing and reproducing. By-products from growing roots and plant residues feed soil organisms. In turn, soil organisms support plant health as they decompose organic matter, cycle nutrients, enhance soil structure and control the populations of soil organisms, both beneficial and harmful (pests and pathogens) in terms of crop productivity.

The living part of soil organic matter includes a wide variety of micro-organisms such as bacteria, viruses, fungi, protozoa and algae. It also includes plant roots, insects, earthworms, and larger animals such as moles, mice and rabbits that spend part of their life in the soil. The living portion represents about 5 per cent of the total soil organic matter. Micro-organisms, earthworms and insects help break down crop residues and manures by ingesting them and mixing them with the minerals in the soil, and in the process recycling energy and plant nutrients.

Sticky substances on the skin of earthworms and those produced by fungi and bacteria help bind particles together. Earthworm casts are also more strongly aggregated (bound together) than the surrounding soil as a result of the mixing of organic matter and soil mineral material, as well as the intestinal mucus of the worm. Thus, the living part of the soil is responsible for keeping air and water available, providing plant nutrients, breaking down pollutants and maintaining the soil structure. The composition of soil organisms depends on the food source (which in turn is season dependent). Therefore, the organisms are neither uniformly distributed through the soil nor uniformly present all year. However, in some cases their biogenic structures remain. Each species and group exists where it can find appropriate food supply, space, nutrients and moisture. Organisms occur wherever organic matter occurs. Therefore, soil organisms are concentrated: around roots, in litter, on humus, on the surface of soil aggregates and in spaces between aggregates. For this reason, they are most prevalent in forested areas and cropping systems that leave a lot of biomass on the surface.

Fig. 4.8 Plate Termites Create Their Own Living Conditions Near Their Preferred Food Sources. Inside

The activity of soil organisms follows seasonal as well as daily patterns. Not all organisms are active at the same time. Most are barely active or even dormant. Availability of food is an important factor that influences the level of activity of soil organisms and thus is related to land use and management (Figure 4.8). Practices that increase numbers and activity of soil organisms include: no tillage or minimal tillage; and the maintenance of plant and annual residues that reduce disturbance of soil organisms and their habitat and provide a food supply.Different groups of organisms can be distinguished in the soil

DECOMPOSITION PROCESS

Fresh residues consist of recently deceased micro-organisms, insects and earthworms, old plant roots, crop residues, and recently added manures.Crop residues contain mainly complex carbon compounds originating from cell walls (cellulose, hemicellulose, etc.). Chains of carbon, with each carbon atom linked to other carbons, form the "backbone" of organic molecules. These carbon chains, with varying amounts of attached oxygen, H, N, P and S, are the basis for both simple sugars and amino acids and more complicated molecules of long carbon chains or rings. Depending on their chemical structure, decomposition is rapid (sugars, starches and proteins), slow (cellulose, fats, waxes and resins) or very slow (lignin).

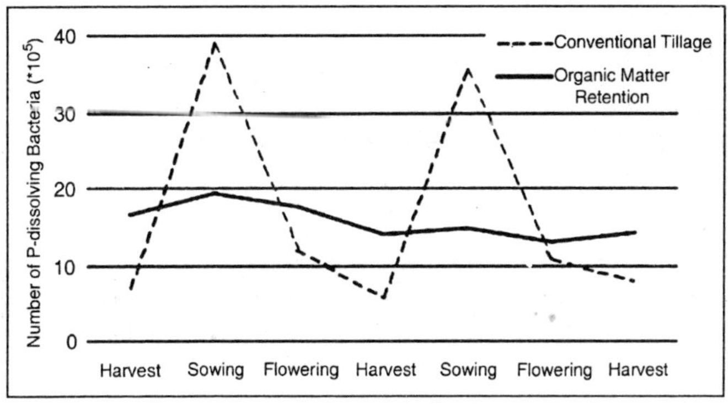

Fig. 4.9 Fluctuations in Microbial Biomass at Different Stages of Crop Development in Conventional Agriculture Compared With Systems With Residue Retention and High Organic Matter Input.

Table. Classification of Soil Organisms

Micro-organisms	Microflora	<5 μm	Bacteria Fungi
	Microfauna	<100 μm	Protozoa Nematodes
Macro-organisms	Meso-organisms	100 μm-2 mm	Springtails Mites
	Macro-organisms	2-20 mm	Earthworms Millipedes

			Woodlice
			Snails and slugs
Plants	Algae	10 μm	
	Roots	> 10 μm	

Table. Essential Functions Performed by Different Members of Soil Organisms (Biota)

Functions	Organisms involved
Maintenance of soil structure	Bioturbating invertebrates and plant roots, mycorrhizae and some other micro-organisms
Regulation of soil hydrological processes	Most bioturbating invertebrates and plant roots
Gas exchange and carbon sequestration (accumulation in soil)	Mostly micro-organisms and plant roots, some C protected in large compact biogenic invertebrate aggregates
Soil detoxification	Mostly micro-organisms
Nutrient cycling	Mostly micro-organisms and plant roots, some soil- and litter-feeding invertebrates
Decomposition of organic matter	Various saprophytic and litter-feeding invertebrates (detritivores), fungi, bacteria, actinomycetes and other micro-organisms
Suppression of pests, parasites and diseases	Plants, mycorrhizae and other fungi, nematodes, bacteria and various other micro-organisms, collembola, earthworms, various predators
Sources of food and medicines	Plant roots, various insects (crickets, beetle larvae, ants, termites), earthworms, vertebrates, micro-organisms and their by-products
Symbiotic and asymbiotic relationships with plants and their roots	Rhizobia, mycorrhizae, actinomycetes, diazotrophic bacteria and various other rhizosphere micro-organisms, ants
Plant growth control (positive and negative)	Direct effects: plant roots, rhizobia, mycorrhizae, actinomycetes, pathogens, phytoparasitic nematodes, rhizophagous insects, plant-growth promoting rhizosphere micro-organisms, biocontrol agents Indirect effects: most soil biota

During the decomposition process, microorganisms convert the carbon structures of fresh residues into transformed carbon products in the soil. There are many different types of organic molecules in soil. Some are simple molecules that have been synthesized directly from plants or other living organisms. These relatively simple chemicals, such as sugars, amino acids, and cellulose are readily consumed by many organisms. For this reason, they do not remain in the soil for a long time. Other chemicals such as resins and waxes also come directly from plants, but are more difficult for soil organisms to break down.

Humus is the result of successive steps in the decomposition of organic matter. Because of the complex structure of humic substances, humus cannot be used by many micro-organisms as an energy source and remains in the soil for a relatively long time.

NON-HUMIC SUBSTANCES: SIGNIFICANCE AND FUNCTION

Non-humic organic molecules are released directly from cells of fresh residues, such as proteins, amino acids, sugars, and starches. This part of soil organic matter is the active, or easily decomposed, fraction. This active fraction is influenced strongly by weather conditions, moisture status of the soil, growth stage of the vegetation, addition of organic residues, and cultural practices, such as tillage. It is the main food supply for various organisms in the soil.

Carbohydrates occur in the soil in three main forms: free sugars in the soil solution, cellulose and hemicellulose; complex polysaccharides; and polymeric molecules of various sizes and shapes that are attached strongly to clay colloids and humic substances. The simple sugars, cellulose and hemicellulose, may constitute 5-25 per cent of the organic matter in most soils, but are easily broken down by micro-organisms.

Polysaccharides (repeating units of sugar-type molecules connected in longer chains) promote better soil structure through their ability to bind inorganic soil particles into stable aggregates. Research indicates that the heavier polysaccharide molecules may be more important in promoting aggregate stability and water infiltration than the lighter molecules. Some sugars may stimulate seed germination and root elongation. Other soil properties affected by polysaccharides include CEC, anion retention and biological activity.

The soil lipids form a very diverse group of materials, of which fats, waxes and resins make up 2-6 per cent of soil organic matter. The significance of lipids arises from the ability of some compounds to act as growth hormones. Others may have a depressing effect on plant growth.

Soil N occurs mainly (> 90 per cent) in organic forms as amino acids, nucleic acids and amino sugars. Small amounts exist in the form of amines, vitamins, pesticides and their degradation products, etc. The rest is present as ammonium (NH_4^-) and is held by the clay minerals.

COMPOUNDS AND FUNCTION OF HUMUS

Humus or humified organic matter is the remaining part of organic matter that has been used and transformed by many different soil organisms. It is a relatively stable component formed by humic substances, including humic acids, fulvic acids, hymatomelanic acids and humins. It is probably the most widely distributed organic carbon-containing material in terrestrial and aquatic environments.

Humus cannot be decomposed readily because of its intimate interactions with soil mineral phases and is chemically too complex to be used by most organisms. It has many functions. One of the most striking characteristics of humic substances is their ability to interact with metal ions, oxides, hydroxides, mineral and organic compounds, including toxic pollutants, to form water-soluble and water-insoluble complexes.

Through the formation of these complexes, humic substances can dissolve, mobilize and transport metals and organics in soils and waters, or accumulate in certain soil horizons. This influences nutrient availability, especially those nutrients present at microconcentrations only. Accumulation of such complexes can contribute to a reduction of toxicity, *e.g.* of aluminium (Al) in acid soils, or the capture of pollutants-herbicides such as Atrazine or pesticides such as Tefluthrin-in the cavities of the humic substances.

Humic and fulvic substances enhance plant growth directly through physiological and nutritional effects. Some of these substances function as natural plant hormones (auxines and gibberillins) and are capable of improving seed germination, root initiation, uptake of plant nutrients and can serve as sources of N, P and S. Indirectly, they may affect plant growth through modifications of physical, chemical and biological properties of the soil, for example, enhanced soil water holding capacity and CEC, and improved tilth and aeration through good soil structure.

About 35-55 per cent of the non-living part of organic matter is humus. It is an important buffer, reducing fluctuations in soil acidity and nutrient availability. Compared with simple organic molecules, humic substances are very complex and large, with high molecular weights. The characteristics of the well-decomposed part of the organic matter, the humus, are very different from those of simple organic molecules. While much is known about their general chemical composition, the relative significance of the various types of humic materials to plant growth is yet to be established.

Humus consists of different humic substances:

- *Fulvic acids:* The fraction of humus that is soluble in water under all pH conditions. Their colour is commonly light yellow to yellow-brown.
- *Humic acids:* The fraction of humus that is soluble in water, except for conditions more acid than pH 2. Common colours are dark brown to black.
- *Humin:* The fraction of humus that is not soluble in water at any pH and that cannot be extracted with a strong base, such as sodium hydroxide (NaOH). Commonly black in colour.

The term acid is used to describe humic materials because humus behaves like weak acids.Fulvic and humic acids are complex mixtures of large molecules. Humic acids are larger than fulvic acids. Research suggests that the different substances are differentiated from each other on the basis of their water solubility. Fulvic acids are produced in the earlier stages of humus formation. The relative amounts of humic and fulvic acids in soils vary with soil type and management practices. The humus of forest soils is characterized by a high content of fulvic acids, while the humus of agricultural and grassland areas contains more humic acids.

NATURAL FACTORS INFLUENCING THE AMOUNT OF ORGANIC MATTER

The transformation and movement of materials within soil organic matter pools is a dynamic process influenced by climate, soil type, vegetation and soil organisms. All these factors operate within a hierarchical spatial scale. Soil organisms are responsible for the decay and cycling of both macronutrients and micronutrients, and their activity affects the structure, tilth and productivity of the soil.

In natural humid and subhumid forest ecosystems without human disturbance, the living and non-living components are in dynamic equilibrium with each other. The litter on the soil surface beneath different canopy layers and high biomass production generally result in high biological activity in the soil and on the soil surface. *Mollison and Slay distinguished the following five mechanisms:*

1. A continuous soil cover of living plants, which together with the soil architecture facilitates the capture and infiltration of rainwater and protects the soil;
2. A litter layer of decomposing leaves or residues providing a continuous energy source for macro- and micro-organisms;
3. The roots of different plants distributed throughout the soil at different depths permit an effective uptake of nutrients and an active interaction with microorganisms;

4. The major period of nutrient release by micro-organisms coincides with the major period of nutrient demand by plants;
5. Nutrients recycled by deep-rooting plants and soil macrofauna and microfauna.

This equilibrium creates almost closed-cycle transfers of nutrients between soil and the vegetation adapted to such site conditions, resulting in almost perfect physical and hydric conditions for plant growth, *i.e.* a cool microclimate, increased evapotranspiration, good rooting conditions with good porosity and sufficient soil moisture. This facilitates water infiltration and prevents erosion and run-off. Thus, it results in clean water in the streams emanating from the area, a relatively smooth variation in streamflow during the year, and recharge of groundwater.

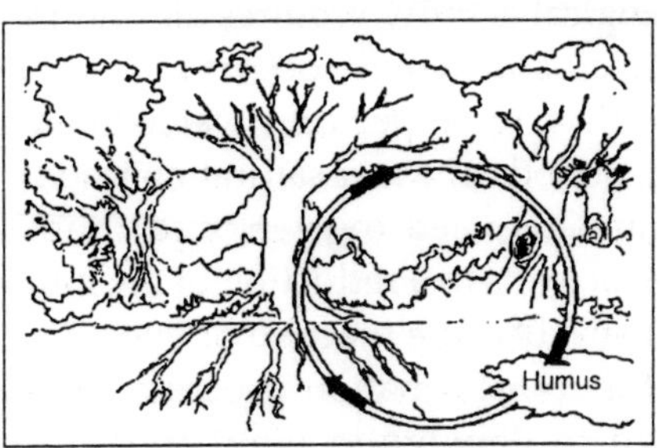

Fig. 4.10 Closed Cycle

TEMPERATURE

Several field studies have shown that temperature is a key factor controlling the rate of decomposition of plant residues. Decomposition normally occurs more rapidly in the tropics than in temperate areas.

Ladd and Amato reported that, despite differences in plant material and climate patterns, the decomposition of leguminous materials in southern Australian sites followed the same pattern as that of ryegrass for sites in Nigeria and the United Kingdom, although the time scales were different. Reaction rates doubled for each increase of 8-9 °C in the mean annual air temperature. The relatively faster rate of decomposition induced by the continuous warmth in the tropics implies that high equilibrium levels of organic matter are difficult to achieve in tropical agro-ecosystems. Hence, large annual rates of organic inputs are needed to maintain an adequate labile soil organic matter pool in cultivated soils. Soils in cooler climates commonly have more organic matter because of slower mineralization (decomposition) rates.

SOIL MOISTURE AND WATER SATURATION

Soil organic matter levels commonly increase as mean annual precipitation increases. Conditions of elevated levels of soil moisture result in greater biomass production, which provides more residues, and thus more potential food for soil biota. Soil biological activity requires air and moisture.

Optimal microbial activity occurs at near "field capacity", which is equivalent to 60-per cent water-filled pore space. In the other hand, periods of water saturation lead to poor aeration. Most soil organisms need oxygen, and thus a reduction of oxygen in the soil leads to a reduction of the mineralization rate as these organisms become inactive or even die.

Some of the transformation processes become anaerobic, which can lead to damage to plant roots caused by waste products or favourable conditions for disease-causing organisms. Continued production and slow decomposition can lead to very large organic matter contents in soils with long periods of water saturation.

With the exception of the hyperhumid regions, the climates of vast areas of the humid, subhumid and semi-arid tropics are characterized by distinct wet and dry seasons. In the wet-dry tropics, large amounts of nitrate often occur in the surface soil during the first part of the rainy season. This accelerated nitrogen mineralization caused by a large increase in microbial activity is the result of the first few rains activating the labile soil organic matter. Farmers who practise "slash and burn" agriculture often choose early planting in order to take advantage of this flush of inorganic N before it is lost through leaching and run-off. In these low-input systems, the amount of nitrate present in the soil during the early part of the rainy season is related closely to the organic matter content of the soil. N availability diminishes during the later part of the rainy season.

SOIL TEXTURE

Soil organic matter tends to increase as the clay content increases. This increase depends on two mechanisms. First, bonds between the surface of clay particles and organic matter retard the decomposition process.

Second, soils with higher clay content increase the potential for aggregate formation. Macroaggregates physically protect organic matter molecules from further mineralization caused by microbial attack.

For example, when earthworm casts and the large soil particles they contain are split by the joint action of several factors (climate, plant growth and other organisms), nutrients are released and made available to other components of soil micro-organisms. Under similar climate conditions, the organic matter content in fine textured (clayey) soils is two to four times that of coarse textured (sandy) soils.

Kaolinite, the main clay mineral in many upland soils in the tropics, has a much smaller specific surface and nutrient exchange capacity than most other clay minerals. Therefore, kaolinitic soils contain considerably fewer clay-humus complexes.

In addition, the unprotected labile humic substances are vulnerable to decomposition under appropriate soil moisture conditions. Thus, high levels of organic matter are difficult to maintain in cultivated kaolinitic soils in the wet-dry tropics, because climate and soil conditions favour rapid decomposition.

In contrast, organic matter can persist as organo-oxide complexes in soils rich in iron and aluminium oxides. Such properties favour the formation of soil microaggregates, typical of many fine-textured, oxide-rich, high base-status soils in the tropics. These soils are known for their low bulk density, high microporosity, and high organic-matter retention under natural vegetation, but also for their high phosphate fixation capacity on the oxides when used for crop production.

Current knowledge suggests that whereas organic matter contributes to the dark colour of Vertisols, it is not considered important in determining either the development, robustness or resilience of structure in these soils. Organic matter levels tend to be low in Vertisols; even as low as 10 g/ kg.

Parent material influences organic matter accumulation not only through its effect on soil texture. Soils developed from inherently rich material, such as basalt, are more fertile than soils formed from granitic material, which contains less mineral nutrients. Moreover, the former experience more organic matter accumulation because of abundant vegetative growth.

TOPOGRAPHY

Organic matter accumulation is often favoured at the bottom of hills. There are two reasons for this accumulation: conditions are wetter than at mid- or upper-slope positions, and organic matter is transported to the lowest point in the landscape through run-off and erosion.

Similarly, soil organic matter levels are higher on northfacing slopes (in the Northern Hemisphere) compared with south-facing slopes (and the other way around in the Southern Hemisphere) because temperatures are lower.

SALINITY AND ACIDITY

Salinity, toxicity and extremes in soil pH (acid or alkaline) result in poor biomass production and, thus in reduced additions of organic matter to the soil. For example, pH affects humus formation in two ways: decomposition, and biomass production. In strongly acid or highly alkaline soils, the growing conditions for micro-organisms are poor, resulting in low levels of biological oxidation of organic matter.

Soil acidity also influences the availability of plant nutrients and thus regulates indirectly biomass production and the available food for soil biota. Fungi are less sensitive than bacteria to acid soil conditions.

VEGETATION AND BIOMASS PRODUCTION

The rate of soil organic matter accumulation depends largely on the quantity and quality of organic matter input. Under tropical conditions, applications of readily degradable materials with low C:N ratios, such as green manure and leguminous cover crops, favour decomposition and a short-term increase in the labile nitrogen pool during the growing season. On the other hand, applications of plant materials with both large C:N ratios and lignin contents such as cereal straw and grasses (Figure 4.11) generally favour nutrient immobilization, organic matter accumulation and humus formation, with increased potential for improved soil structure development.

Plant constituents such as lignin and other polyphenols retard decomposition. In an experiment in southern Nigeria to compare management effects on soil organic matter accumulation, a three-year fallow with Guinea grass (*Panicum maximum*), which has a high lignin content, maintained a carbon level comparable to that under forest fallow.

However, fallowing with leguminous species such as pigeon pea (*Cajanus cajan*) caused a significant decline in soil total C (Juo and Lal, 1977).

Palm and Sanchez reported that both the decomposition rate and the N-release patterns of three tropical legumes (*Inga edulis, Cajanus cajan,* and *Erythrina* spp.) were related to the amount of polyphenol compounds such as lignin in the leaf. *Erythrina* leaves had the lowest concentrations of polyphenols and the fastest decomposition rate of the three species studied.

Root turnover also constitutes an important addition of humus into the soil, and consequently it is important for carbon sequestration. In forests, most organic matter is added as superficial litter. However, in grassland ecosystems, up to two-thirds of organic matter is added through the decay of roots.

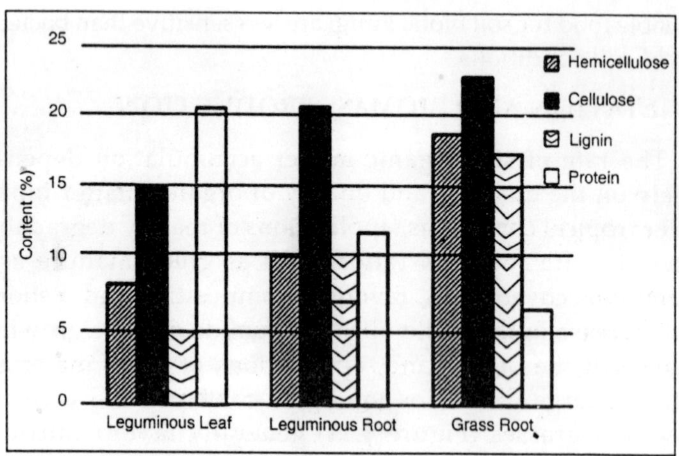

Fig. 4.11 Composition of Leaves and Roots of Leguminous and Grass Species.

Chapter 5

Influencing Amount of Organic Matter

HUMAN INTERVENTIONS THAT INFLUENCE SOIL ORGANIC MATTER

Various types of human activity decrease soil organic matter contents and biological activity. However, increasing the organic matter content of soils or even maintaining good levels requires a sustained effort that includes returning organic materials to soils and rotations with high-residue crops and deep- or dense-rooting crops. It is especially difficult to raise the organic matter content of soils that are well aerated, such as coarse sands, and soils in warm-hot and arid regions because the added materials decompose rapidly. Soil organic matter levels can be maintained with less organic residue in finetextured soils in cold temperate and moist-wet regions with restricted aeration.

PRACTICES THAT DECREASE SOIL ORGANIC MATTER

Any form of human intervention influences the activity of soil organisms and thus the equilibrium of the system. Management practices that alter the living and nutrient conditions of soil organisms, such as repetitive tillage or burning of vegetation, result in a degradation of their microenvironments.

In turn, this results in a reduction of soil biota, both in biomass and diversity. Where there are no longer organisms to decompose soil organic matter and bind soil particles, the soil structure is damaged easily by rain, wind and sun. This can lead to rainwater run-off and soil erosion, removing the potential food for organisms, *i.e.* the organic matter of the topsoil. Therefore, soil biota are the most important property of the soil, and "when devoid of its biota, the uppermost layer of earth ceases to be soil".

Fig. 5.1 Open Cycle System

The factors leading to reduction in soil organic matter in an open cycle system (Figure 5.1) can be grouped as factors that result in:

- A decrease in biomass production;
- A decrease in organic matter supply;
- Increased decomposition rates.

DECREASE IN BIOMASS PRODUCTION

REPLACEMENT OF PERENNIAL VEGETATION

A consequence of clearing forest for agriculture is the disappearance of the litter layer, with a consequent reduction in the numbers and variety of soil organisms. While many temperate forest species appear to adapt well to grassland, the effects of deforestation in the tropics appear to be more marked. Studies have shown that as soil biodiversity declines, adapted species may take over from the indigenous species and the composition may change drastically. Soil macrofaunal biomass and population density fell to 6 and 17 per cent, respectively, in cultivated plots, compared with primary forest in Peruvian Amazonia. In Suriname, the number of animals per square metre has fallen to 36 per cent and the diversity of species has fallen to 28 per cent compared with primary forest. The indigenous species have largely disappeared, but adapted species have been available for recolonization. The composition of the macrofaunal community has changed drastically (Figure 5.2).

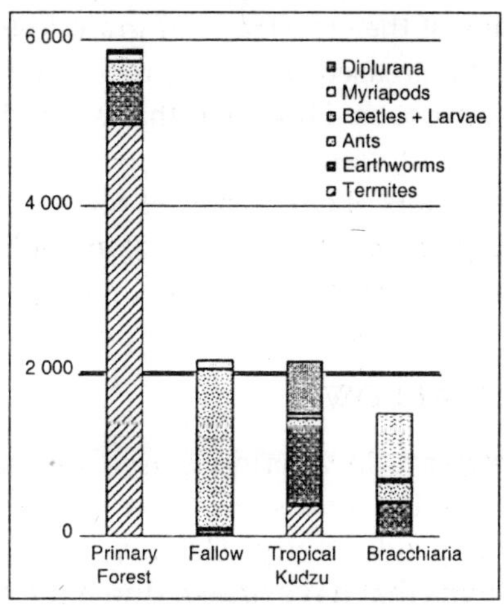

Fig. 5.2 Composition of Soil Macrofauna Under Primary Forest, Fallow, Kudzu and Grass Vegetation

REPLACEMENT OF MIXED VEGETATION WITH MONOCULTURE OF CROPS AND PASTURES

The simplification of vegetation and the disappearance of the litter layer under grassland and monocrop production systems lead to a decrease in faunal diversity. Although root systems (especially of grasses) can be extensive and explore vast areas of soil, the root exudates from one single crop will attract only a few different microbial species. This in turn will affect the predator diversity. The more opportunistic pathogen species will be able to acquire space near the crop and cause harm. Continuous cultivation and grazing also leads to compaction of soil layers, which in turn affects the circulation of air. Anaerobic conditions in the soil stimulate the growth of different micro-organisms, resulting in more pathogenic organisms.

HIGH HARVEST INDEX

One of the consequences of the green revolution was the replacement of indigenous varieties of species with high-yielding varieties (HYVs). These HYVs often produce more grain and less straw, compared with locally developed varieties; the harvest index of the crop (ratio of grain to total plant mass aboveground) is increased. From a production point of view, this is a logical approach. However, this is less desirable from a conservation point of view.

Reduced amounts of crop residues remain after harvest for soil cover and organic matter, or for grazing of livestock (which results in manure). Moreover, where animals graze the residues, even less remains for conservation purposes.

USE OF BARE FALLOW

Traditionally, a fallow period is used after a period of crop production to give the land some "rest" and to regenerate its original state of productivity. Usually, this is necessary in production systems that have drawn down the nutrient supply and altered the soil biota significantly, such as in slash-and-burn systems or conventional tillage systems.

Some farmers use bare fallow to regenerate their lands. However, apart from spontaneous weed growth, this means there is no energy source for the soil biota present on the land. Instead of recovering the soil food web, the soil organic matter is degraded further and the lack of cover can result in severe erosion and run-off when the rains start after the dry season.

DECREASE IN ORGANIC MATTER SUPPLY

BURNING OF NATURAL VEGETATION AND CROP RESIDUES

The burning of maize, rice and other crop residues in the field is a common practice. Residues are usually burned to help control insects or diseases or to make fieldwork easier in the following season. Burning destroys the litter layer and so diminishes the amount of organic matter returned to the soil.

The organisms that inhabit the surface soil and litter layer are also eliminated. For future decomposition to take place, energy has to be invested first in rebuilding the microbial community before plant nutrients can be released. Similarly, fallow lands and bush are burned before cultivation. This provides a rapid supply of P to stimulate seed germination. However, the associated loss of nutrients, organic matter and soil biological activity has severe long-term consequences.

OVERGRAZING

Cows, draught animals and small ruminants graze on communal grazing areas and on roadsides, stream banks and other public land. Overgrazing destroys the most palatable and useful species in the plant mixture and reduces the density of the plant cover, thereby increasing the erosion hazard and reducing the nutritive value and the carrying capacity of the land.

REMOVAL OF CROP RESIDUES

Many farmers remove residues from the field for use as animal feed and bedding or to make compost. Later, these residues return to contribute to soil fertility as manures or composts. However, residues are sometimes removed from the field and not returned. This removal of plant material impoverishes the soil as it is no longer possible to recycle the plant nutrients present in the residues.

INCREASED DECOMPOSITION RATES

TILLAGE PRACTICES

Tillage is one of the major practices that reduces the organic matter level in the soil. Each time the soil is tilled, it is aerated. As the decomposition of organic matter and the liberation of C are aerobic processes, the oxygen stimulates or speeds up the action of soil microbes, which feed on organic matter.

This means that:

- When ploughed, the residues are incorporated in the soil together with air and come into contact with many micro-organisms, which accelerates the carbon cycle. The decomposition is faster, resulting in the formation of less stable humus and an increased liberation of CO_2 to the atmosphere, and thus a reduction in organic matter.

- The residues on the soil surface slow the carbon cycle because they are exposed to fewer micro-organisms and thus wane more slowly, resulting in the production of humus (which is more stable), and liberating less CO_2 to the atmosphere.

Table. Tillage Induced Flush of Decomposition of Organic Matter.

Type of Tillage	Organic Matter Lost in 19 Days (kg/ha)
Mouldboard plough + disc harrow (2x)	4 300
Mouldboard plough	2 230
Disc harrow	1 840
Chisel plough	1 720
Direct seeding	860

In terms of short-term organic matter loss, the more a soil is tilled, the more the organic matter is broken down. There are also longer-term losses, attributed to repeated, annual cultivation. Cropping systems that return little residue to the soil accelerate this decline. Many modern cropping systems combine frequent tillage with small amounts of residue, with resultant reductions in the organic matter content of many soils. Historically, manure application (from farm livestock) was common, and it was a dynamic way of maintaining organic matter levels despite repeated cultivation and low residue returns to the soil. Increased on-farm mechanization has reduced livestock numbers, so this source of organic material has been reduced considerably. Organic matter production and conservation is affected dramatically by conventional tillage, which not only decreases soil organic matter but also increases the potential for erosion by wind and water.

The impact occurs in many ways:

- Ploughing leaves no residues on the soil surface to lessen the impact of rain.
- Ploughing reduces the quantity of food sources for earthworms and disturbs their burrows and living space, hence populations of certain species decrease drastically. Moreover, reduction of earthworm numbers reduces their impact, through burrowing, in increasing porosity and aeration (particularly continuous macropores) and lowers their ability to bury and incorporate plant residues, which facilitates rapid decomposition of organic matter.
- Tillage by repeated hoeing or discing smoothes the surface and destroys natural soil aggregates and channels that connect the surface with the subsoil, leaving the soil susceptible to erosion. Old root channels and earthworm holes are eliminated, as are the cracks between natural aggregates. The large pores, the ones destroyed by conventional tillage practices, are necessary to conduct water into the soil during rainfall.
- The development of a plough pan or hoe pan, a layer of compacted soil resulting from smearing action at the bottom of the plough or hoe, may retard both root penetration and water infiltration.
- Ploughing or discing under dry conditions exacerbates the pulverization of the soil, causing the soil surface to crust more easily, leading to greater water run-off and erosion. This is exacerbated by reduced soil surface roughness, which leaves few depressions for temporary storage of water during intense storms.
- Increased run-off during rainstorms may also increase the possibility of drought stress later in the season, because water that runs off the field does not infiltrate into the soil to remain available to plants.

In some circumstances, imbalances of certain soil organisms can disrupt soil structure and processes, *e.g.* certain earthworm species in rice fields or pastures.

DRAINAGE

Decomposition of organic matter occurs more slowly in poorly aerated soils, where oxygen is limiting or absent, compared with well-aerated soils. For this reason, organic matter accumulates in wet soil environments. Soil drainage is determined strongly by topography-soils in depressions at the bottom of hills tend to remain wet for extended periods of time because they receive water (and sediments) from upslope. Soils may also have a layer in the subsoil that inhibits drainage, again exacerbating waterlogging and reduction in organic matter decomposition. In a permanently waterlogged soil, one of the major structural parts of plants, lignin, does not decompose at all. The ultimate consequence of extremely wet or swampy conditions is the development of organic (peat or muck) soils, with organic matter contents of more than 30 per cent. Where soils are drained artificially for agricultural or other uses, the soil organic matter decomposes rapidly.

FERTILIZER AND PESTICIDE USE

Initially, the use of fertilizer and pesticides enhances crop development and thus production of biomass (especially important on depleted soils). However, the use of some fertilizers, especially N fertilizers, and pesticides can boost micro-organism activity and thus decomposition of organic matter. The chemicals provide the microorganisms with easy-to-use N components. This is especially important where the C: N ratio of the soil organic matter is high and thus decomposition is slowed by a lack of N.

PRACTICES THAT INCREASE SOIL ORGANIC MATTER

Increased concern about the environmental and economic impacts of conventional crop production has stimulated interest in alternative systems. Central to such systems is the need to promote and maintain soil biological processes and minimize fossil fuel inputs in the form of fertilizers, pesticides and mechanical cultivation. All activities aimed at the increase of organic matter in the soil help in creating a new equilibrium in the agro-ecosystem.

For a system of natural resource management to be balanced, and thus sustainable, it must be able to withstand sharp climatic fluctuations, and to evolve steadily in response to social changes and changes in the costs and availability of inputs of land, labour and knowledge. The more diverse and complex an agricultural system is, the more stable and sustainable it will be in the face of unpredictable vagaries of climate and market. Thus, annual crops, woody perennials and nonwoody perennials may be combined in various ways with livestock or trees, or both, in what are now commonly called agrosilvipastoral systems. Different approaches are required for different soil and climate conditions. However, the activities will be based on the same principle: increasing biomass production in order to build active organic matter. Active organic matter provides habitat and food for beneficial soil organisms that help build soil structure and porosity, provide nutrients to plants, and improve the water holding capacity of the soil.

Several cases have demonstrated that it is possible to restore organic matter levels in the soil (Figure 5.3). Activities that promote the accumulation and supply of organic matter, such as the use of cover crops and refraining from burning, and those that reduce decomposition rates, such as reduced and zero tillage, lead to an increase in the organic matter content in the soil.

Ways to increase organic matter contents of soils:

- Compost
- Cover crops/green manure crops
- Crop rotation
- Perennial forage crops
- Zero or reduced tillage
- Agroforestry

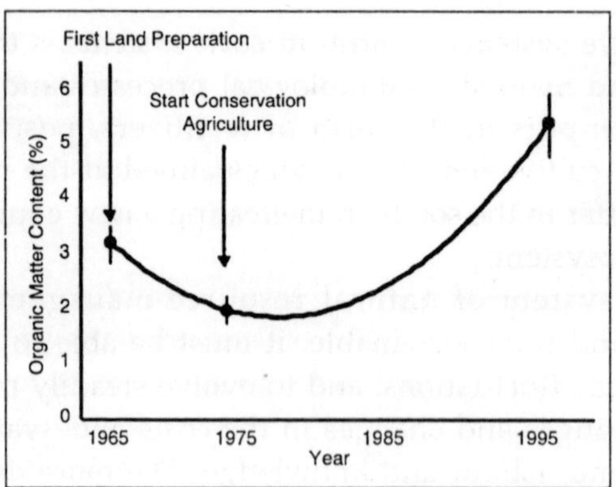

Fig. 5.3 Evaluation of The Organic Matter Content of a Soil in Paraná

INCREASED BIOMASS PRODUCTION

INCREASED WATER AVAILABILITY FOR PLANTS: WATER HARVESTING AND IRRIGATION

In dry conditions, water may be provided through irrigation or water harvesting. The increased water availability enhances biomass production, soil biological activity and plant residues and roots that provide organic matter.The concept of water harvesting includes various technologies for run-off management and utilization. It involves capture of run-off (in some cases through treating the upstream capture area), and its concentration on a runon area for use by a specific crop (annual or perennial) in order to enhance crop growth and yields, or its collection and storage for supplementary irrigation or domestic or livestock purposes.

The objective of designing a water harvesting system is to obtain the best ratio of the area yielding run-off to either the area where run-off is being directed or the capacity of the storage structure (volume of water collected). In this way, the water captured for crop production during run-off periods can be stored either directly in the soil for subsequent use by plants or in small farm reservoirs or collection tanks. This aids stabilization of crop production by enhancing soil moisture availability or allowing irrigation during a dry period within the rainy season or by extending crop production into the dry season. Some factors to be considered regarding these run-off farming systems and reservoirs include: site selection, watershed size and condition, rainfall distribution and run-off, and water requirements of crops. Where a minimum water depth of about 1 m can be maintained in a reservoir, fish can be raised to provide additional food. Numerous water harvesting systems have been developed over the centuries, especially in arid areas. The principle of collecting run-off for crop production is also inherent to many other soil and water conservation technologies that apply the concept of run-off and runon areas at a microwatershed level, such as negarims, trapezoidal or "eyebrow" bunds and tied ridges.

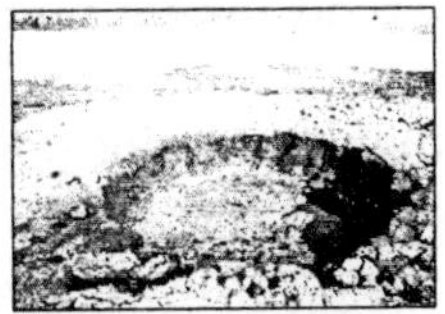

Fig.5.4 Plate9

BALANCED FERTILIZATION

Where the supply of nutrients in the soil is ample, crops are more likely to grow well and produce large amounts of biomass. Fertilizers are needed in those cases where nutrients in the soil are lacking and cannot produce healthy crops and sufficient biomass. Most soils in sub-Saharan Africa (SSA) are deficient in P. P is required not only for plant growth but also for N fixation.

Unbalanced fertilization, for example mainly with N, may result in more weed competition, higher pest incidence and loss of quality of the product. Unbalanced fertilization eventually leads to unhealthy plants. Therefore, fertilizers should be applied in sufficient quantities and in balanced proportions. The efficiency of fertilizer use will be high where the organic matter content of the soil is also high.

In very poor or depleted soils, crops use fertilizer applications inefficiently. When soil organic matter levels are restored, fertilizer can help maintain the revolving fund of nutrients in the soil by increasing crop yields and, consequently, the amount of residues returned to the soil.

COVER CROPS

Growing cover crops is one of the best practices for improving organic matter levels and, hence, soil quality.

The benefits of growing cover crops include:

- They prevent erosion by anchoring soil and lessening the impact of raindrops.
- They add plant material to the soil for organic matter replenishment.
- Some, *e.g.* rye, bind excess nutrients in the soil and prevent leaching.
- Some, especially leguminous species, *e.g.* hairy vetch, fix N in the soil for future use.
- Most provide habitat for beneficial insects and other organisms.
- They moderate soil temperatures and, hence, protect soil organisms.

A range of crops can be used as vegetative cover, *e.g.* grains, legumes and oil crops. All have the potential to provide great benefit to the soil. However, some crops emphasize certain benefits; a useful consideration when planning a rotation scheme. It is important to start the first years with (cover) crops that cover the surface with a large amount of residues that decompose slowly (because of the high C:N ratio). Grasses and cereals are most appropriate for this stage, also because of their intensive rooting system, which improves the soil structure rapidly. In the following years, when soil health has begun to improve, legumes can be incorporated in the rotation. Leguminous crops enrich the soil with N and their residues decompose rapidly because of their low C:N ratio. Later, when the system is stabilized, it is possible to include cover crops with an economic function, *e.g.* livestock fodder. The selection of cover crops should depend on the presence of high levels of lignin and phenolic acids. These give the residues a higher resistance to decomposition and thus result in soil protection for a longer period and the production of more stable

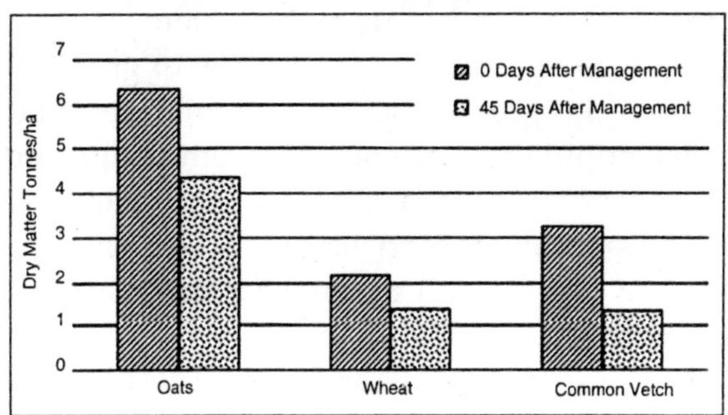

Fig. 5.5 Reduction of Dry Matter of Different Cover Crops

Another determining factor in the dynamics of residue composition is the biochemical composition of the residues. Depending on species, their chemical components and the time and way of managing them, there will be differences in decomposition rates (Figure 5.5). The grain species (oats and wheat) show more resistance than common vetch (legume) to decomposition. The latter has a lower C:N ratio and a lower lignin content and is thus subject to a rapid decomposition.

Agricultural production systems in which residues are left on the soil surface, such as direct seeding and the use of cover crops, stimulate the development and activity of soil fauna at many levels. The term green manure is often used to indicate the same plant species that are used as cover crops. However, green manure refers specifically to a crop in the rotation grown for incorporation of the non-decomposed vegetative matter in the soil.

While this practice is used specifically to add organic matter, this is not the most effective use of organic matter (especially in hot climates) for two reasons:

1. Mechanical disturbance of the soil should be avoided as much as possible.
2. When biomass is incorporated in the soil all at one

time, there is a short period of high microbial activity in decomposing the material. This results in the sudden release of a large quantity of nutrients that cannot be captured by the seedlings of the following crop and is thus lost from the system.

In general, the greater the production of green manure or crop biomass, the greater is the microbial, mesofauna and macrofauna population of the soil-from fungi and micro-organisms to earthworms and termites. The dynamics of surface residue decomposition depend *inter alia* on the activity of micro-organisms and also on soil mesofauna and macrofauna. The macrofauna consists mainly of earthworms, beetles, termites, ants, millipedes, spiders, snails and slugs. These organisms help integrate the residues into the soil and improve soil structure, porosity, water infiltration, and through-flow through the creation of burrows, ingestion and secretions. The natural incorporation of cover-crop and weed residues from the soil surface to deeper layers in the soil by soil macrofauna is a slow process. The activity of microorganisms is regulated by the activity of the macrofauna, because the latter provide them with food and air through their bioturbation activities.

In this way, nutrients are released slowly and can provide the crop with nutrients over a longer period. At the same time, the soil is covered for a long time by the residues and is protected against the impact of rain and sun.

IMPROVED VEGETATIVE STANDS

In many places, low plant densities limit crop yields. Wide plant spacing is often practised as "a way to return power to the soil" or "to give the soil some rest", but in reality it is an indicator that the soil is impoverished. Plant spacing is usually determined by farmers in relation to soil fertility and available water or expected rainfall (unless standard recommendations are enforced by extension). This means that plants are often spaced widely on depleted soils in arid and semi-arid regions with a view to ensuring an adequate provision of plant nutrients and water for all plants.

However, it is important to maintain the recommended plant spacing in order to optimize biomass production and rooting density and, hence, organic matter for food, moisture retention and habitat for soil organisms. Once the crop is established, reduced sunlight between closer crop rows may also reduce regrowth of weeds.

Planting pits achieve fast rehabilitation of severely degraded land, especially in a semi-arid climate where a short fallow period of natural grass growth (2-6 years after 2-3 years under crops) cannot be expected to maintain or restore the land's agricultural productivity.

An example of the rapid restoration of productivity of degraded land is an indigenous method in the Sahel region called "zaï" During the dry season, farmers dig out pits 15 cm deep and 40 cm in diameter every 80 cm, tossing the earth downhill. The dry desert Harmattan wind blows various organic residues into the excavated pits. The organic materials are consumed quickly by termites, which excavate tunnels through the crusted surface, allowing the first rains to soak down deep, out of danger of direct evaporation.

Two weeks before the onset of the rains, farmers spread one or two handfuls of dry dung (1- 2.5 tonnes/ha) in the bottom of the pits and cover it with earth to prevent the rains from eroding away the organic matter.

Millet is sown into the pits at the onset of the rainy season. As the first rains wash over the surface crust (of the degraded land), the basins capture this run-off (enough to soak a pocket of soil up to 1 m in depth). The sown seeds germinate, break up the slaked surface crust and send roots down to the deeper stores of both water and nutrients (recycled by the termites).

At harvest time, stalks are cut at a height of 1 m and left in situ to reduce wind-speed and trap windborne organic matter. In the second year, the farmer either digs new basins between the first ones and dresses them with manure, or pulls up the stubble and sows again in the old basins. Stubble clumps laid between basins are in turn used as a food source by termites. Planting pits are a way of increasing biomass production and crop yields on severely degraded land in semi-arid conditions. Rainfall is concentrated near the plants, and soil faunal activity and organic matter accumulation are concentrated in the planting pits. Planting pits have been introduced successfully in Zambia as a conservation practice for smallholder farmers, who do not have fertilizers or tractor services available to them.

Fig. 5.6 Plate Half-Moons Around Newly Planted Acacia Seedlings Catch and Retain RainWater.

AGROFORESTRY AND ALLEY CROPPING

Agroforestry is a collective name for land-use systems where woody perennials (trees, shrubs, palms, etc.) are integrated in the farming system. Alley cropping is an agroforestry system in which crops are grown between rows of planted woody shrubs or trees. These are pruned during the cropping season to provide green manure and to minimize shading of crops.

Agroforestry covers a wide range of systems combining food crops, forestry and pasture species in different ways (agrosilviculture, silvipasture, agrosilvipasture and multipurpose forest production). There are two different approaches to agroforestry.

One uses agricultural crops or pasture as a transitional means of utilizing the land until forest plantations are fully established. The other is to integrate trees and shrubs permanently into the crop or animal production system, to the benefit of both crop production and land resource protection. Thus, agroforestry encompasses many traditional land-use systems such as home gardens, shifting cultivation and bush fallow systems.

Examples of agroforestry systems worldwide

Poro (*Erythrina poeppigiana*) has been grown extensively in coffee plantations in Costa Rica for shade, soil enrichment, live mulching and live fences.

Albizzia spp. have been used in tea plantations in many Asian countries. In Indonesia, leucaena (*Leucaena leucocephala*) has been planted as contour hedges on hillsides for erosion control, soil improvement and green mulch. It is estimated that some 20 000 ha of undulating land have been converted to these systems.

In West Africa and Rwanda, many farmers use trees, fruit trees, bushes and grasses planted with agricultural crops on their farms. Many coconut plantations in the Caribbean are partly planted with bananas or used as pastures. Some small farms in Jamaica plant coconuts, banana and citrus together.

Alley cropping can be considered an improved bush fallow system. Small trees or shrubs are planted in cropland in rows, preferably along the contour (even where east-west orientation of the rows may minimize shading of crops). The optimal spacing between rows depends on: slope; soil type and its susceptibility to erosion; rainfall; crop species; and the soil and crop management system.

Besides adding organic matter to the system, perennial trees and shrubs recycle plant nutrients from deeper soil layers through their rooting system. Through litter and pruning, these can be used again by annual crops. Probably the most important contribution of perennials in a production system lies in the fact that throughout the whole year their roots excrete root exudates and decaying root cells, which in turn are used as an energy source by soil microorganisms. The food web in the soil is maintained, even during dry seasons when no annual crops are grown. The result is that soil biota are in place to provide the crop with nutrients at the beginning of the next cropping season. Direct seeding is the easiest and cheapest way of establishing hedgerows around fields or in the fields (alleys). However, emerging seedlings may not be able to compete with weeds without additional care. Therefore, starting plant growth in a nursery and transplanting may be necessary for some species. Other species may be established by cuttings. With good establishment, the plants will be better able to withstand both dry spells and browsing by livestock. Crucial to a successful establishment of the hedgerow is that the selected plants should be tall enough to outgrow the weeds at the time of the first crop harvest.

Fig. 5.7 Plate

During the cropping season, hedgerow pruning is needed in order to avoid shading of the crop. The timing, frequency and extent of pruning depend on the species used and the season. As a general rule, the lower the hedgerows and the taller the crop, the less frequently is pruning required. Fastgrowing plants such as *Leucaena leucocephala* and *Gliricidia sepium* may require pruning every six weeks during the cropping season. They are often pruned to a height of about 50 cm. Care must be exercised as too frequent pruning can result in tree dieback.

The integration of trees and woody shrubs into the cropping system offers additional uses and many benefits, as mentioned by farmers using the Quezungual system in Honduras. However, farmers with short-term land tenure may not be interested in these benefits. Furthermore, the plantation of trees sometimes has an effect on the land tenure status; therefore tenants may not be allowed to establish trees on agricultural land. Agroforestry systems can also inhibit mechanization and may need increased labour inputs, especially for hedgerow pruning.

Farmers' perceptions of the Quezungual system: benefits and disadvantages

The Quezungual system has many benefits according to local farmers:

- Improved soil moisture conservation, which permits a good development of the crop even during the dry spells of 2-4 weeks halfway through the rainy season;
- Production of fuelwood and fruits from the trees and shrubs;
- Agricultural production is greater than in traditionally managed plots;
- Plots with the Quezungual system can be cultivated for longer periods than under the slash-andburn system;
- Timber trees can be cut after about 7 years and used for construction and/or sold;
- The mulch obtained through the pruning of the trees and shrubs protects the soil surface from the impact of rain showers, thus there is less soil erosion (even during the heavy rains produced by hurricane "Mitch" in 1998 there was little soil erosion);
- Minimal labour is required to establish and maintain the Quezungual system;
- The soil becomes more fertile and the effect of fertilizers on production improves;
- The workability of the soil improves because the soil becomes softer, hence less labour is needed during sowing;
- The Quezungual system provides shade for farmers while they work the plot;
- Harvested products, such as beans and maize, can be dried by hanging them over the tree trunks;
- Cattle can feed on the residues after maize and sorghum harvests;
- Mulch cover reduces the incidence of disease in the bean crop;
- The presence of trees and shrubs in the plot attracts animals and insects, *e.g.* birds and butterflies.

Disadvantages mentioned by the farmers are:

- In the first year, the production is the same as or slightly less than grain production obtained with the traditional system;
- In the early years of implementation, the incidence of slugs in the bean crop is greater.
- Too much soil cover can impede seed germination;
- The shade of the Quezungual system can result in a higher incidence of disease during intense rainfall periods.

The hedgerow species have to be selected carefully in order to avoid negative impacts on crop production because of the complex relationships (competition for light, water and nutrients, allelopathy, occurrence of pest and diseases, etc.) that are inherent to agroforestry systems. Many farmers may consider the hedgerows as not useful, especially where their positive effects are not secure or visible. Where livestock are allowed to graze freely, it can be difficult to establish hedgerows without taking special measures to protect the young plants.

Fencing or control of grazing animals may require collective efforts and agreement by the local community. The increased labour requirement, the reduced cropped area and the difficulty of mechanization may make alley cropping uneconomic unless the hedgerow species produce direct benefits such as fruits, fuelwood or poles/timber for construction purposes (in addition to the nutrient recycling and erosion control effects).

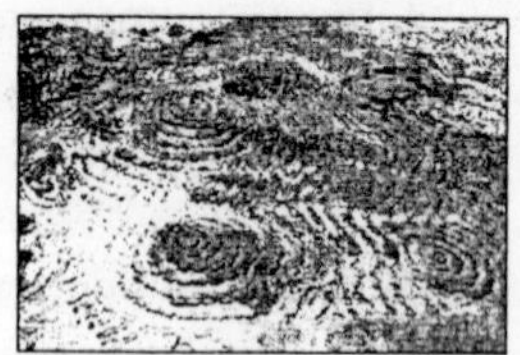

Fig. 5.8 Plate

REFORESTATION AND AFFORESTATION

Afforestation means the establishment of a forest on land that has not grown trees recently. It can serve two principal soil and water conservation purposes: protection of erosion-prone areas, and revegetation and rehabilitation of degraded land. Afforestation is specifically used to provide protective cover in vulnerable, steep and mountainous areas. Afforestation helps to replenish timber resources and provide fuelwood and fodder.

The establishment of a forest cover under good management is an effective means of increasing organic matter production. However, the land must have the productive capacity to support an appropriate forest type, which differs according to climate, soil, slope and the specific purpose of the forest (timber production, livestock grazing, etc.). Therefore, the choice of species and the selection of an appropriate site are of particular importance for successful afforestation.

The procurement of adequate quantities of good quality seed of the species and provenances (adapted varieties) required is a prerequisite for any afforestation effort. However, it is often diffi2cult to find suitable and reliable sources of such seeds. A number of species require special pre-treatment of the seed or seedling in order to achieve satisfactory germination and uniform stand. Such treatment may consist of soaking the seed in water for varying lengths of time, alternate soaking and drying, scarifying or chipping the seed coat to render it permeable to water, plunging the seed into boiling water or even boiling it for a short time. Some tree seedlings may need a mycorrhizal treatment when planted in soils that are deprived of associated mycorrhizae species as well as rhizobium species (*e.g. Casuarina* in Senegalese sandy soils).

The aim is to ensure that good numbers of plants germinate and that germination after sowing is both rapid and uniform. Afforestation can be achieved by direct sowing or replanting young plants from a nursery. The main advantage of direct sowing is the reduced cost.

However, this is usually much less reliable and is only justified where:

- Seed is plentiful and cheap;
- Adequate germination under field conditions can be relied on;

- The seedlings send down a deep tap root rapidly and are able to withstand adverse climatic conditions in the time after germination;
- The rate of growth is sufficiently fast to make a prolonged period of tending and weeding unnecessary.

REGENERATION OF NATURAL VEGETATION

Regeneration of natural grasslands and forest areas increases biomass production and improves the plant species diversity, resulting in more diverse soil biota and other associated beneficial organisms. Natural regeneration may be more reliable where land is not very productive. In some cases, natural regeneration of a given area may lead to the infestation of plots by weeds. Increasingly, natural vegetation is being recognized for its multipurpose benefits, for example, fuelwood, fibre, biocontrol (*e.g.* neem) and medicinal species, as well as restoration of soil fertility (*Acacia albida* and other leguminous species) and habitats for various beneficial species (pollinators and natural enemies) as well as wildlife.

INCREASED ORGANIC MATTER SUPPLY

PROTECTION FROM FIRE

Burning affects organic matter recycling significantly. Fire destroys almost all organic materials on the land surface except for tree trunks and large branches. In addition, the surface soil is sterilized, loses part of its organic matter, the population of soil microfauna and macrofauna is reduced, and no ready-to-use organic matter is available for rapid restoration of the populations. However, this practice is widely used (*e.g.* in Africa) in order to enhance pasture regrowth for livestock (using residual P), to control pests and diseases, and even to catch small animals for food.

A specific and difficult case is the burning of sugar cane before harvesting. It has both a technical dimension (CO_2 and greenhouse gas emissions, mechanization of harvest, sugar content, etc.) and a social dimension (manual cutting, source of survival resources for poor/landless workers). The damage depends on fire intensity, which is a function of vegetation type and climate conditions and frequency. The costs and benefits of burning and the methods to minimize harmful effects need to be identified with local populations.

CROP RESIDUE MANAGEMENT

In systems where crop residues are managed well, they:

- Add soil organic matter, which improves the quality of the seedbed and increases the water infiltration and retention capacity of the soil, buffers the pH and facilitates the availability of nutrients;
- Sequester (store) C in the soil;
- Provide nutrients for soil biological activity and plant uptake;
- Capture the rainfall on the surface and thus increase infiltration and the soil moisture content;
- Provide a cover to protect the soil from being eroded;
- Reduce evaporation and avoid desiccation from the soil surface.

Depending on the nature of the following crop, decisions are made as to whether the residues should be distributed evenly over the field or left intact, *e.g.* where climbing cover crops (*e.g.* mucuna) use the maize stalks as a trellis.

An even distribution of residues:

- Provides homogenous temperature and humidity conditions at sowing time;
- Facilitates even sowing, germination and emergence;
- Minimizes the development of pests and diseases; and
- Reduces the emergence of weeds through allelopathic effects.

The most appropriate method for managing crop residues depends on the purpose of the crop residues and the experience and equipment available to the farmer. Where the aim is to maintain a mulch over the soil for as long as possible, the biomass is best managed using a knife roller, chain or sledge in order to break it down but not kill it. Where the decomposition process should commence immediately in order to release nutrients, the residues should be slashed or mown and some N applied because dry residues have a high C:N ratio. However, in order to avoid nitrate emission, urea should not be broadcast on the surface but injected where possible.

Fig. 5.9 Plate

UTILIZING FORAGE BY GRAZING RATHER THAN BY HARVESTING

In many places, there is competition for the use of crop residues that can be used as fodder, for roofing, artisan handicrafts, etc. Where residues are to be used for animal feed, either the animals graze the residues directly, or they are stall- or kraal-fed. Removal of the residues from the field can lead to a considerable loss of organic matter where animal manure is not returned to the field. By controlled grazing, the animal manure is returned in the field without a high labour input.

The experience of Guaymango, El Salvador, demonstrates that it is possible to achieve successful integration of crop and livestock components without creating competition in the allocation of crop residues. The amount of residues produced by the system is enough to serve both as soil cover and as fodder for livestock, mainly because of the use of local sorghum varieties (instead of HYVs) that have a high straw/grain ratio. As farmers value crop residues as soil cover, a fodder market has developed where grazing rights, number of cattle and duration of grazing are traded.

In the northern zone of the United Republic of Tanzania, farmers have found a compromise between using the residues for grazing or soil cover, albeit one that is rather labour intensive. They separate the palatable and non-palatable parts of the crop residues. They use the non-palatable parts to cover the soil and act as food for soil organisms, while they feed the palatable parts to cattle and goats that are kept close to the homestead.

INTEGRATED PEST MANAGEMENT

As with balanced fertilization, proper pest and disease management results in healthy crops. Healthy crops produce optimal biomass, which is necessary for organic matter production in the soil. Diversified cropping and mixed crop-livestock systems enhance biological control of pests and diseases through species interactions. Through integrated production and pest management farmers learn how to maintain a healthy environment for their crops.

They learn to examine their crops regularly in order to observe ratios of pests to natural enemies (beneficial predators) and cases of damage, and on that basis to make decisions as to whether it is necessary to use natural treatments (using local products such as neem or tobacco) or chemical treatments and the required applications.

Fig. 5.10 Plate

APPLYING ANIMAL MANURE OR OTHER CARBON-RICH WASTES

Any application of animal manure, slurry or other carbon-rich wastes, such as coffee-berry pulp, improves the organic matter content of the soil. In some cases, it is better to allow a period of decomposition before application to the field. Any addition of carbon-rich compounds immobilizes available N in the soil temporarily, as micro-organisms need both C and N for their growth and development. Animal manure is usually rich in N, so N immobilization is minimal. Where straw makes up part of the manure, a decomposition period avoids N immobilization in the field.

COMPOST

Composting is a technology for recycling organic materials in order to achieve enhanced agricultural production. Biological and chemical processes accelerate the rate of decomposition and transform organic materials into a more stable humus form for application to the soil. Composting proceeds under controlled conditions in compost heaps and pits.

Compost heaps should have a minimum size of 1 m^3 and are suitable for more humid environments where there is potential for watering the compost. Compost pits should be no deeper than 70 cm and should be underlain with rough material for good aeration of the compost. Pits are suitable for drier environments where the compost may desiccate. Dry composting relies on covering the compost with soil and creating an anaerobic environment. However, this is a slower process than the more usual moist aerobic process. The ratio of C to N in the compost pile is important for optimizing microbial activity. Thus, a mixture of soft, green and brown, tougher material is used. Ash and phosphate rock are often added to accelerate the process.

Composting can complement certain crop rotations and agroforestry systems. It can be used efficiently in planting pits and nurseries. It is very similar in composition to soil organic matter. It breaks down slowly in the soil and is very good at improving the physical condition of the soil (whereas manure and sludge may break down fairly quickly, releasing a flush of nutrients for plant growth). In many circumstances, it takes time to rejuvenate a poor soil using these practices because the amount of organic material being added is small relative to the mineral proportion of the soil.

Successful composting depends upon the sufficient availability of organic materials, water, manure and "cheap" labour. Where these inputs are guaranteed, composting can be an important method of sustainable and productive agriculture. It has ameliorative effects on soil fertility and physical, chemical and biological soil properties. Well-made compost contains all the nutrients needed by plants. It can be used to maintain and improve soil fertility as well as to regenerate degraded soil. However, materials for compost production may be in short supply and the technology demands high labour inputs for proper compost production and application. Therefore, compost application may be restricted to certain crops and limited application areas, *e.g.* vegetable production in home gardens.

MULCH OR PERMANENT SOIL COVER

One way to improve the condition of the soil is to mulch the area requiring amelioration. Mulches are materials placed on the soil surface to protect it against raindrop impact and erosion, and to enhance its fertility. Crop residue mulching is a system of maintaining a protective cover of vegetative residues such as straw, maize stalks, palm fronds and stubble on the soil surface. The system is particularly valuable where a satisfactory plant cover cannot be established rapidly when erosion risk is greatest. Mulching adds organic matter to the soil, reduces weed growth, and virtually eliminates erosion during the period when the ground is covered with mulch.

There are two principal mulching systems:

1. *In situ* mulching systems-plant residues remain where they fall on the ground;
2. Cut-and-carry mulching systems-plant residues are brought from elsewhere and used as mulch.

Crop residue mulching has numerous positive effects on crop production. However, it may require a change in existing cropping practices. For example, farmers may conventionally burn crop residues instead of returning them to the soil. In situ mulching depends on the design of appropriate cropping systems and crop rotations, which have to be integrated with the farming system. The greater labour demands of cut-and-carry systems represent a major constraint. Mulch may be more relevant in home gardens or for valuable horticulture crops than in less intensive farming systems. Mulch affects the soil life. Holland and Coleman have demonstrated that litter placement on the soil surface (as opposed to incorporation with ploughing) increased the ratio of fungi to bacteria-the reason being that fungi have a higher carbon assimilation efficiency than bacteria. In addition, it encourages bioturbating (mixing) effects of macrofauna that pull the materials into surface layers of the soil.

DECREASED DECOMPOSITION RATES

REDUCED OR ZERO TILLAGE

Repetitive tillage degrades the soil structure and its potential to hold moisture, reduces the amount of organic matter in the soil, breaks up aggregates, and reduces the population of soil fauna such as earthworms that contribute to nutrient cycling and soil structure. Avoiding mechanical soil disturbance implies growing crops without mechanical seedbed preparation or soil disturbance since the harvest of the previous crop. The term zero tillage is used for this practice synonymously with terms such as no-till farming, no tillage, direct drilling, and direct seeding.

Compared with conventional tillage, reduced or zero tillage has two advantages with respect to soil organic matter. Conventional tillage stimulates the heterotrophic microbiological activity through soil aeration, resulting in increased mineralization rate. Through breakdown of soil structure, it decreases upward and downward movements of soil fauna, such as earthworms, which are largely responsible for "humus" production through the ingestion of fresh residues. Reduced or zero tillage regulates heterotrophic microbiological activity because the pore atmosphere is richer in CO_2/O_2, and facilitates the activity of the "humifiers". Mulching in the highlands of northern Thailand Why certain crops receive mulch and others do not Mulching provides a particular benefit to the cultivated crop. Mulching is practised for various cash crops for specific reasons. Onion and garlic are mulched mainly to control weeds (early hand weêding would be difficult without damaging the crop). The mulch is also important to keep the soil moist and cool as these crops are usually grown during the dry season under irrigation. Mulch is also applied under flowers and strawberries, mainly to protect the fragile and valuable products from becoming soiled. Mulching saves labour. Mulching is often seen in maize fields, before as well as after crop establishment. Maize can compete reasonably well with weeds. Therefore, some farmers plant maize without tillage in a mulch of weeds previously killed with herbicides-a system that is less labour-demanding than a tillage operation. Because maize is planted with large spacings, it generally requires less rigorous weeding. Weeding is often done by slashing, and the weed residues are left on the ground.

Fig. 5.11 Plate *Mulching in The Highlands of Northern Thailand.*

Tillage has become the most common method to control weeds. However, mulching is a more environmentally sound practice than tillage for weed control. The loose soil that results from tilling has less structure than before; the appearance is deceptive. Subsequent traffic or heavy rain soon packs this loosened soil, not only negating the expensive cultivation that produced the loose soil but also culminating in a degraded environment for water entry, seed germination and root growth.

Further cultivation is then required to re-loosen the soil; more expense with the same outcome-subsequent repacking and degraded soil structure. This is a typical "downward spiral" of conventional agriculture. Moreover, tillage when the soil is too moist or too dry leads to compaction or pulverization of soil; but farmers may not have the option to wait for optimal conditions. Severe, accelerated soil erosion and the high costs in terms of labour and energy associated with plough-based methods of seedbed preparation have led to the widespread adoption of no- or zero-tillage systems for cropping in temperate and tropical climates. In no-tillage systems, the crop is sown into a soil left undisturbed since the harvest of the previous crop. Crop residue mulch is maintained and anchored firmly to the ground. Weed control relies on mechanical slashing or cover crops.

In reduced- or zero-tillage systems, soil fauna resume their bioturbating activities gradually. These loosen the soil and mix the soil components (also known as biotillage). The additional benefit of the increased soil organic matter and burrowing is the creation of a stable and porous soil structure without expensive, time-consuming and potentially degrading cultivations. In zero-tillage systems, the action of soil macrofauna gradually incorporate cover crop and weed residues from the soil surface down into the soil. The activity of microorganisms is also regulated by the activity of the macrofauna, which provide them with food and air through their burrows. In this way, nutrients are released slowly and can provide the following crop with nutrients.

Several authors have demonstrated that some crop rotations and zero tillage favour *Bradyrhizobia* populations, nodulation and thus N fixation and yield. Figure 5.12 indicates a 200-300-per cent increase in population size of root nodule bacteria in a zero-tillage system compared with conventional tillage. The presence of soybean in the crop rotation resulted in a fivefold to tenfold increase in population size of the same bacteria compared with cropping systems without soybean.

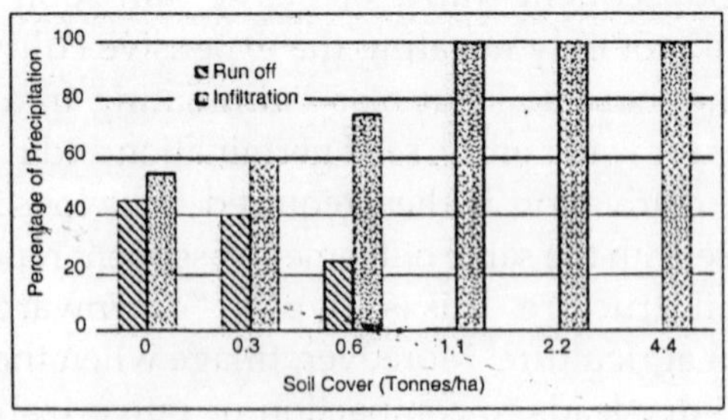

Fig. 5.12 Population Size of Root Nodule Bacteria With Different Crop Rotations

Note: S = soybean; W = wheat; M = maize.

Strictly speaking, the term zero tillage applies to methods involving no soil disturbance whatsoever, a condition that may be difficult to achieve. Broadcasting of seed is one way of applying zero tillage. The seed is broadcast over the previous crop residues and, where necessary, the residues are shaken to ensure that the seed falls on the soil surface.

In direct drilling, seeds such as maize, sorghum, soybean, wheat and barley are sown directly into shallow furrows cut into the previous crop residues. Weeds are controlled mechanically with a knife, which knocks down the plants and breaks their stems, or chemically with herbicides.

Traditional practices such as the burning of crop residues may inhibit the introduction of no-tillage systems. In many situations, a conflict exists between leaving crop residues on the surface or feeding them to livestock in the dry season when there is a shortage of fodder.

Mechanical soil disturbance also includes soil compaction through wheel impact of machinery, especially important in large-scale mechanized agriculture, *e.g.* plantations (sugar cane) or biannual crops (cotton). In a zero-tillage farming system, consideration must be given to reducing both the random placement of tyres/wheels in fields as well as the potential for compaction from animal hooves. Pietola, Horn and Yli-Halla reported the destructive effect of cattle trampling on the soil structure.

Proffitt, Bendotti and McGarry demonstrated the almost total loss of soil porosity in the soil surface as a result of trampling by sheep. There is a belief that draught animals cause less land degradation than tractors. However, there are reports of soil compaction on smallholder farming enterprises in both Malawi and Bangladesh.

The hooves of draught animals and the shearing effect of ploughs or hand hoes, which are used repeatedly at a constant depth, can cause severe compacted layers. Grazing animals should be removed from zero-till fields in moist-wet soil conditions as the compaction risk is greatest at these times.

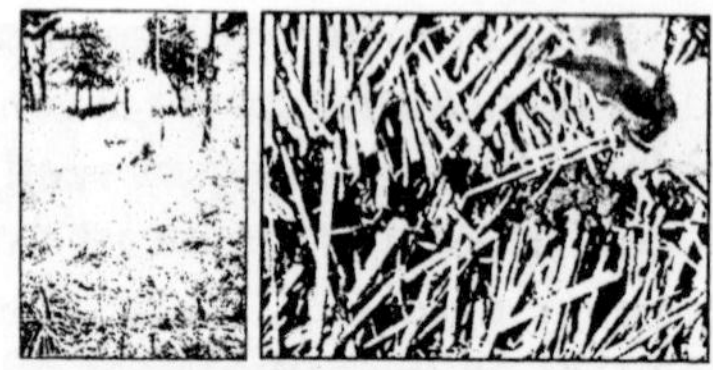

Fig. 5.13 Plate Maize Seedling Directly Drilled in Residues of Wheat.

Controlled traffic, where the wheels of all in-field equipment follow permanent, defined tracks, ensures that compaction is restricted to specific known areas. Alternatively, flotation tyres (low ground-pressure tyres) should be fitted to all large tractors, harvesters, in-field grain bins, etc. in order to reduce their compacting potential.

Recent research has demonstrated the devastating effects of compaction from wheel impact on the occurrence and survival of eartworms. Earthworm incidence was greater under controlled traffic than under wheeled traffic. Figure 5.13 shows the immediate effects of wheeling and tillage on the earthworm population. It appears that wheeling has the most detrimental effect on earthworm survival and that where wheeling is followed by tillage the survival rate is much greater. This may indicate that earthworms are able to survive an initial compaction in the field as long as it is relieved immediately. Where it is not, earthworms are inmobilized and unable to find air and nutrients.

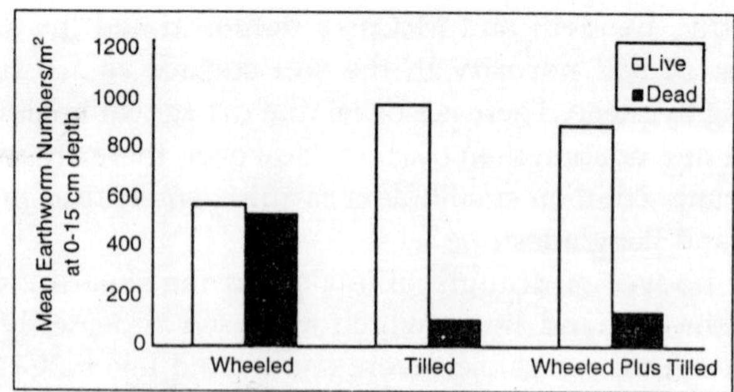

Fig. 5.14 Live and Dead Earthworm Numbers Per Square Metre at 0-15 cm of Soil Depth Sampled Immediately After Treatment.

Chapter 6

Drought-Resistant Soil

EFFECT OF SOIL ORGANIC MATTER ON SOIL PROPERTIES

Organic matter affects both the chemical and physical properties of the soil and its overall health. Properties influenced by organic matter include: soil structure; moisture holding capacity; diversity and activity of soil organisms, both those that are beneficial and harmful to crop production; and nutrient availability. It also influences the effects of chemical amendments, fertilizers, pesticides and herbicides.

INEFFICIENT USE OF RAINWATER

Drylands may have low crop yields not only because rainfall is irregular or insufficient, but also because significant proportions of rainfall, up to 40 per cent, may disappear as run-off.

This poor utilization of rainfall is partly the result of natural phenomena (relief, slope, rainfall intensity), but also of inadequate land management practices (*i.e.* burning of crop residues, excessive tillage, eliminating hedges, etc.) that reduce organic matter levels, destroy soil structure, eliminate beneficial soil fauna and do not favour water infiltration. However, water "lost" as run-off for one farmer is not lost for other water users downstream as it is used for recharging groundwater and river flows.

Where rainfall lands on the soil surface, a fraction infiltrates into the soil to replenish the soil water or flows through to recharge the groundwater. Another fraction may run off as overland flow and the remaining fraction evaporates back into the atmosphere directly from unprotected soil surfaces and from plant leaves.

The above-mentioned processes do not occur at the same moment, but some are instantaneous (run-off), taking place during a rainfall event, while others are continuous (evaporation and transpiration).

To minimize the impact of drought, soil needs to capture the rainwater that falls on it, store as much of that water as possible for future plant use, and allow for plant roots to penetrate and proliferate. Problems with or constraints on one or several of these conditions cause soil moisture to be one of the main limiting factors for crop growth.

The capacity of soil to retain and release water depends on a broad range of factors such as soil texture, soil depth, soil architecture (physical structure including pores), organic matter content and biological activity. However, appropriate soil management can improve this capacity.

Practices that increase soil moisture content can be categorized in three groups:

1. Those that increase water infiltration;
2. Those that manage soil evaporation; and
3. Those that increase soil moisture storage capacities. All three are related to soil organic matter.

In order to create a drought-resistant soil, it is necessary to understand the most important factors influencing soil moisture.

INCREASED SOIL MOISTURE

Organic matter influences the physical conditions of a soil in several ways. Plant residues that cover the soil surface protect the soil from sealing and crusting by raindrop impact, thereby enhancing rainwater infiltration and reducing run-off. Surface infiltration depends on a number of factors including aggregation and stability, pore continuity and stability, the existence of cracks, and the soil surface condition. Increased organic matter contributes indirectly to soil porosity (via increased soil faunal activity).

Fresh organic matter stimulates the activity of macrofauna such as earthworms, which create burrows lined with the glue-like secretion from their bodies and are intermittently filled with worm cast material.

The proportion of rainwater that infiltrates into the soil depends on the amount of soil cover provided (Figure 6.1). The figure 6.2 shows that on bare soils (cover = 0 tonnes/ha) run-off and thus soil erosion is greater than when the soil is protected with mulch.

Crop residues left on the soil surface lead to improved soil aggregation and porosity, and an increase in the number of macropores, and thus to greater infiltration rates.

Increased levels of organic matter and associated soil fauna lead to greater pore space with the immediate result that water infiltrates more readily and can be held in the soil The improved pore space is a consequence of the bioturbating activities of earthworms and other macro-organisms and channels left in the soil by decayed plant roots.

On a site in southern Brazil, rainwater infiltration increased from 20 mm/h under conventional tillage to 45 mm/h under no tillage. Over a long period, improved organic matter promoted good soil structure and macroporosity. Water infiltrates easily, similar to forest soils (Figure 6.2).

The consequence of increased water infiltration combined with a higher organic matter content is increased soil storage of water (Figure 6.3). Organic matter contributes to the stability of soil aggregates and pores through the bonding or adhesion properties of organic materials, such as bacterial waste products, organic gels, fungal hyphae and worm secretions and casts.

Moreover, organic matter intimately mixed with mineral soil materials has a considerable influence in increasing moisture holding capacity. Especially in the topsoil, where the organic matter content is greater, more water can be stored.

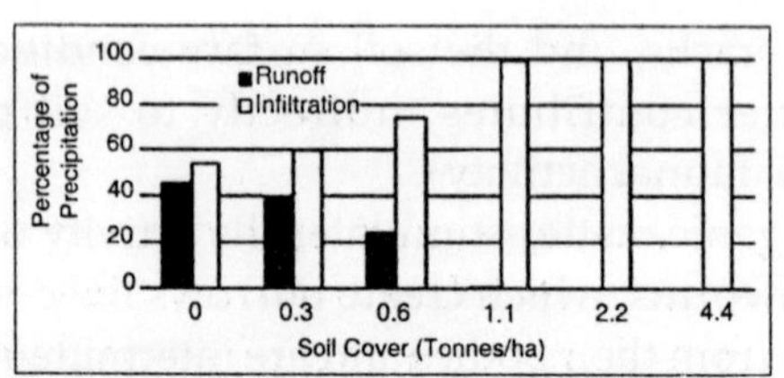

Fig. 6.1 Effect of Amount of Soil Cover on Rainwater Run-off and Infiltration

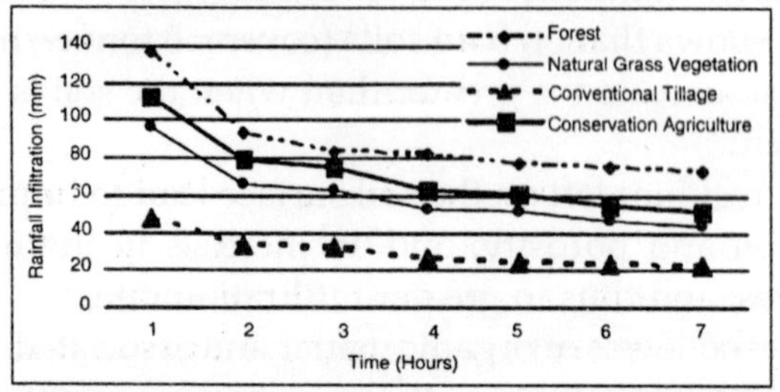

Fig. 6.2 Water Infiltration Under Different Types of Management

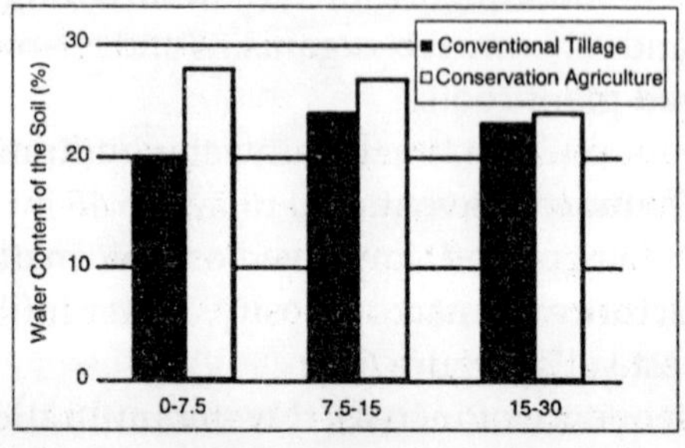

Fig. 6.3 Quantity of Water Stored in The Soil Under Conventional Tillage and Conservation Agriculture

The quality of the crop residues, in particular their chemical composition, determines the effect on soil structure and aggregation. Blair *et al* report a rapid breakdown of medic (*Medicago truncatula*) and rice (*Oryza sativa*) straw residues resulting in a rapid increase in soil aggregate stability through the release of many soilbinding components. As these compounds undergo further breakdown, they will be lost from the system resulting in a decline in soil aggregate stability over time.

The slow release of soil-binding agents from flemingia (*Flemingia macrophylla*) residues resulted in a slower but more sustained increase in the stability of soil aggregates. This indicates that continual release of soil-binding compounds from plant residues is necessary for continual increases in soil aggregate stability to occur.

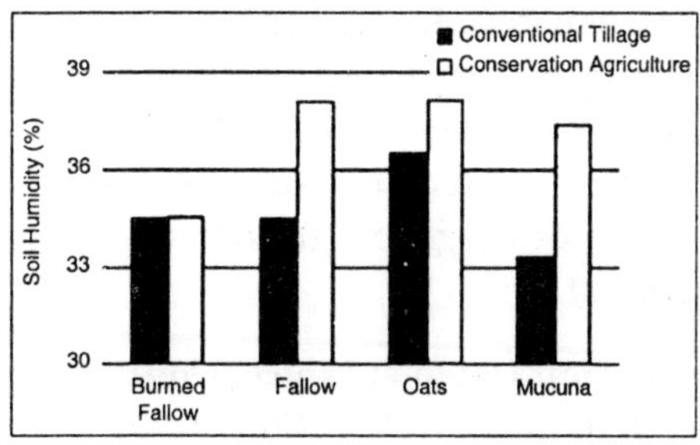

Fig. 6.4 Effect of Different Soil Covers on in-Soil Storage of Water

Elliot and Lynch showed that soil aggregation is caused primarily by polysaccharide production in situations where residues have a low N content. There is a strong relationship between soil carbon content and aggregate size. An increase in soil carbon content led to a 134-per cent increase in aggregates of more than 2 mm and a 38-per cent decrease in aggregates of less than 0.25 mm. The active fraction of soil C is the primary factor controlling aggregate breakdown. In addition, although they do not live long and new ones replace them annually, the hyphae of actinomycetes and fungi play an important role in connecting soil particles. Gupta and Germida showed a reduction in soil macroaggregates correlated strongly with a decline in fungal hyphae after six years of continuous cultivation. The in-soil storage of water depends not only on the type of land preparation but also on the type of cover or previous vegetation on the soil. Figure 6.4 indicates the effect of burning vegetation on the amount of water stored in the soil. Conserving fallow vegetation as a cover on the soil surface, and thus reducing evaporation, results in 4 per cent more water in the soil. This is roughly equivalent to 8 mm of additional rainfall. This amount of extra water can make the difference between wilting and survival of a crop during temporary dry periods. A study conducted in 1999 in Guatemala, Honduras and Nicaragua to evaluate the resilience of agro-ecosystems showed that 3-15 per cent more water was stored in the soil under more ecologically sound practices. Unger showed that high wheat-residue levels resulted in increased storage of fallow precipitation, which subsequently produced higher sorghum grain yields. High residue levels of 8-12 tonnes/ha resulted in about 80-90 mm more stored soil water at planting and about 2.0 tonnes/ha more of sorghum grain yield compared to no residue management.

Table. Average Soil Depth at Which Moisture Starts, and Difference in Moisture Stored

Country	Agro-ecologically soundpractices cm	Conventiona practices cm	Difference (%)
Honduras	9.98	10.28	2.9
Guatemala	2.44	2.99	15.0
Nicaragua	15.81	17.80	11.2

The addition of organic matter to the soil usually increases the water holding capacity of the soil. This is because the addition of organic matter increases the number of micropores and macropores in the soil either by "gluing" soil particles together or by creating favourable living conditions for soil organisms. Certain types of soil organic matter can hold up to 20 times their weight in water. Hudson showed that for each 1-per cent increase in soil organic matter, the available water holding capacity in the soil increased by 3.7 per cent. Soil water is held by adhesive and cohesive forces within the soil and an increase in pore space will lead to an increase in water holding capacity of the soil. As a consequence, less irrigation water is needed to irrigate the same crop (Table).

Table. Economy of Irrigation Water Through Soil Cover, The Brazilian Cerrados

Country	Agroecologically sound practices (cm)	Conventional practices	Difference (%)
Honduras	9.98	10.28	2.9
Guatemala	2.44	2.99	15.0
Nicaragua	15.81	17.80	11.2

REDUCED SOIL EROSION AND IMPROVED WATER QUALITY

The less the soil is covered with vegetation, mulches, crop residues, etc., the more the soil is exposed to the impact of raindrops. When a raindrop hits bare soil, the energy of the velocity detaches individual soil particles from soil clods. These particles can clog surface pores and form many thin, rather impermeable layers of sediment at the surface, referred to as surface crusts. They can range from a few millimetres to 1 cm or more; and they are usually made up of sandy or silty particles. These surface crusts hinder the passage of rainwater into the profile, with the consequence that run-off increases. This breaking down of soil aggregates by raindrops into smaller particles depends on the stability of the aggregates, which largely depends on the organic matter content.

Increased soil cover can result in reduced soil erosion rates close to the regeneration rate of the soil or even lower, as reported by Debarba and Amado for an oats and vetch/maize cropping system.

Soil erosion fills surface water reservoirs with sediment, reducing their water storage capacity. Sedimentation also reduces the buffering and filtering capacity of wetlands and the flood-control capacity of floodplains. Sediment in surface water increases wear and tear in hydroelectric installations and pumps, resulting in greater maintenance costs and more frequent replacement of turbines. Sediments can also reach the sea, harming fish, shellfish and coral. Eroded soil contains fertilizers, pesticides and herbicides; all sources of potentially harmful off-site impacts. When the soil is protected with mulch, more water infiltrates into the soil rather than running off the surface. This causes streams to be fed more by subsurface flow rather than by surface run-off. The consequence is that the surface water is cleaner and resembles groundwater more closely compared with areas where erosion and run-off predominate. Greater infiltration should reduce flooding by increased water storage in soil and slow release to streams. Increased infiltration also improves groundwater recharge, thus increasing well supplies.

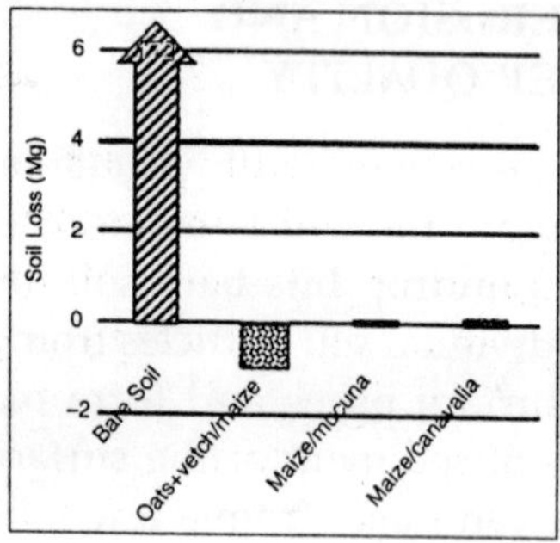

Fig. 6.5 Soil Loss Due to Water Erosion For Different Maize Cropping Systems

Note: Corrected with soil regeneration = 1.7 tonnes/ha/year.

Bassi reported significant reductions in water turbidity and sediment concentration over a period of ten years in different catchment areas in southern Brazil. The reductions varied between 50 and 80 per cent depending on locally predominant soil types. These reductions were caused by increases in the incidence of planting perennial crops (banana and pasture) on hillsides, thereby decreasing erosion potential.

Total sediment loss decreased by 16 per cent and the cost of fertilizers declined by 21 per cent; an indication of the previous loss of fertilizers with the eroded soil. Guimarães, Buaski and Masquieto illustrate the same effect for one specific catchment. The catchment area of Rio do Campo, Paraná, provides 80 per cent of the water supply for Campo Mourão, a city with an urban population of 357 000. In the period 1982-1999, a drastic reduction in water turbidity was measured. Sediment and dissolved organic matter in surface water have to be removed from drinking-water supplies. Reduced erosion, and hence fewer soil particles in suspension, lead to lower costs for water treatment. Data from Chapecó, Brazil, indicate that the quantity of aluminium sulphate used for flocculating suspended solids fell by 46 per cent in five years. Where water is chlorinated to kill disease organisms, the chlorine reacts with dissolved organic matter to form trihalomethane (THM) compounds such as chloroform. THMs are suspected of causing cancers. Reductions in run-off and erosion should lead to reduced formation of THMs during the chlorination process.

Erosion may also have long-lasting secondary consequences through effects on plant growth and litter input. If erosion suppresses productivity, thereby limiting replenishment of organic matter, the amount of organic matter may spiral downwards in the long term. Soil cover protects the soil against the impact of raindrops, prevents the loss of water from the soil through evaporation, and also protects the soil from the heating effect of the sun. Soil temperature influences the absorption of water and nutrients by plants, seed germination and root development, as well as soil microbial activity and crusting and hardening of the soil.

Roots absorb more water at higher soil temperatures up to a maximum of 35 °C. Higher temperatures restrict water absorption. Soil temperatures that are too high are a major constraint on crop production in many parts of the tropics. Maximum temperatures exceeding 40 °C at 5 cm depth and 50 °C at 1 cm depth are commonly observed in tilled soil during the growing season, sometimes with extremes of up to 70 °C. Such high temperatures have an adverse effect not only on seedling establishment and crop growth but also on the growth and development of the micro-organism population. The ideal rootzone temperature for germination and seedling growth ranges from 25 to 35 °C. Experiments have shown that temperatures exceeding 35 °C reduce the development of maize seedlings drastically and that temperatures exceeding 40 °C can reduce germination of soybean seed to almost nil. Mulching with crop residues or cover crops regulates soil temperature. The soil cover reflects a large part of solar energy back into the atmosphere, and thus reduces the temperature of the soil surface. This results in a lower maximum soil temperature in mulched compared with unmulched soil (Figure 6.6) and in reduced fluctuations.

Fig. 6.6 Plate Run-off and Soil Loss Immediately After a Rainstorm, Naisi Catchment. Zomba Mountain, Malawi.

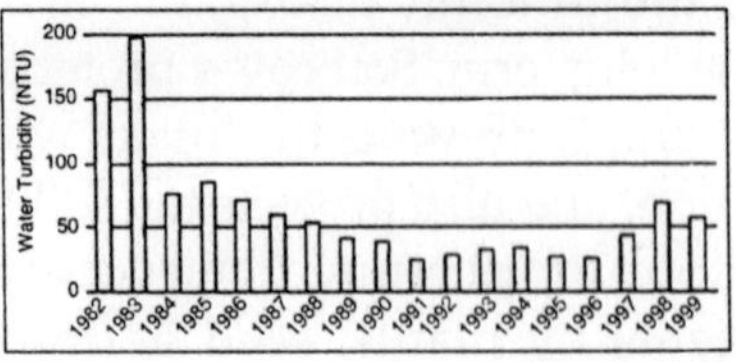

Fig. 6.7 Development of Water Turbidity Rates in The Catchment Area of Rio do Campo, Paraná,

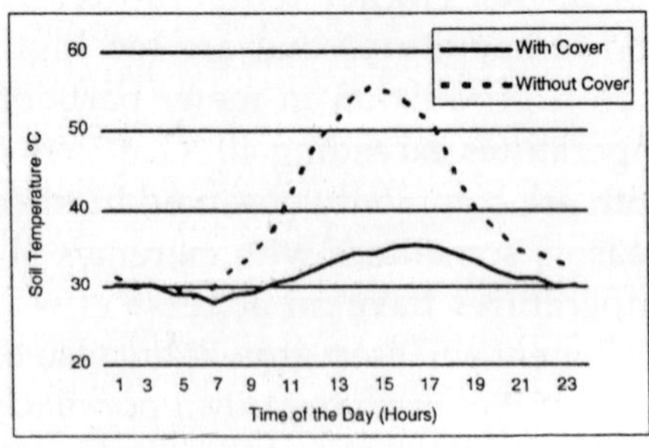

Fig. 6.8 Temperature Fluctuations at a Soil Depth of 3 cm in a Cotton Crop With and Without a Soil Cover of Mucuna

KEY FACTORS IN SUSTAINED FOOD PRODUCTION

INCREASED PLANT PRODUCTIVITY

Plant productivity is linked closely to organic matter. Consequently, landscapes with variable organic matter usually show variations in productivity. Plants growing in well-aerated soils are less stressed by drought or excess water. In soils with less compaction, plant roots can penetrate and flourish more readily. High organic matter increases productivity and, in turn, high productivity increases organic matter.

INCREASED FERTILIZER EFFICIENCY

The two major soil fertility constraints of the West African savannah and in the subhumid and semi-arid regions of SSA are low inherent nutrient reserve and rapid acidification under continuous cultivation as a consequence of low buffering or cation exchange capacity. Generally, these constraints are tackled by applying chemical fertilizers and lime. However, the application of inorganic fertilizers on depleted soils often fails to provide the expected benefits. This is basically because of low organic matter and low biological activity in the soil.

The chemical and nutritional benefits of organic matter are related to the cycling of plant nutrients and the ability of the soil to supply nutrients for plant growth. Organic matter retains plant nutrients and prevents them leaching to deeper soil layers. Microorganisms are responsible for the mineralization and immobilization of N, P and S through the decomposition of organic matter. Thus, they contribute to the gradual and continuous liberation of plant nutrients. Available nutrients that are not taken up by the plants are retained by soil organisms. In organic-matter depleted soils, these nutrients would be lost from the system through leaching and run-off.

Phosphate fixation and unavailability is a major soil fertility constraint in acid soils containing large amounts of free iron and aluminium oxides. In comparing the P-sorption capacity of surface and subsurface soil samples, Uehara and Gilman provided indirect evidence that soil organic matter can reduce the P-sorption capacity of such soils. This implies that for high P-fixing soils, *i.e.* oxide-rich soils derived from volcanic and ferro-magnesian rocks, management systems that are capable of accumulating and maintaining greater amounts of calcium-saturated soil organic matter in the surface horizon would increase P availability from both organic and fertilizer sources.

Weak acids, such as the organic acids in humus, do not relinquish their hydrogen (H) easily. H is part of the humus carboxyl (-COOH) under acidic conditions. When a soil is limed and the acidity decreases, there is a greater tendency for the H+ to be removed from humic acids and to react with hydroxyl (OH-) to form water. The carboxyl groups on the humus develop negative charge as the positively charged H is removed. When the pH of a soil is increased, the release of H from carboxyl groups helps to buffer the increase in pH and at the same time creates the CEC (negative charge). With an increase in organic matter, the soil recovers its natural buffer capacity; this means an increase in pH in acid soils.

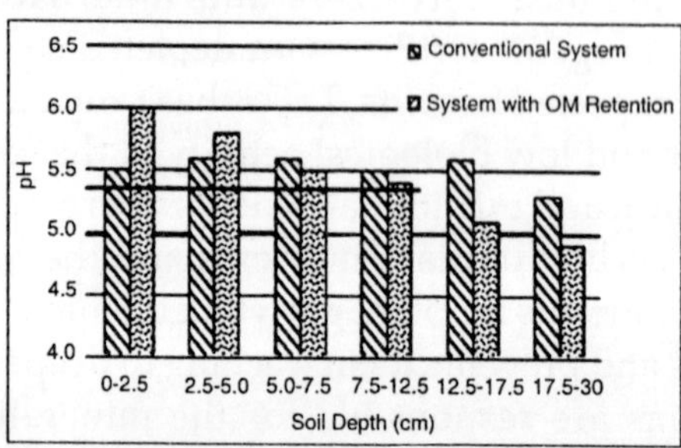

Fig. 6.9 Effect of Organic Matter on Soil pH

Table. Incidence of Lime in The Soil Profile Under Different Soil Covers Over The Same Period After Surface Application

Cover	Soildepth (cm)
Bare soil	0-7
Black oats	0-20
Oil radish	0-22

CEC is linked closely to the organic matter content of the soil. It increases gradually with time where organic residues are retained, first in the topsoil and later also at greater depth. Crovetto reported an increase in CEC of 136 per cent (from 11 to 26 meq/100 g of soil) as a consequence of humus increase in the topsoil after 20 years of residue retention.

To overcome acidity, lime is usually incorporated in the soil. However, organic matter on the soil surface favours the transport of calcium carbonate (lime) to deeper soil layers after surface application (Table).

The crop residues release organic acids that cause the lime to penetrate deeper into the profile much more rapidly than when applied on bare soil. Thus, it is no longer necessary to mix lime intensively into the soil, which is appropriate for farming systems based on reduced or zero tillage.

REDUCED WATERLOGGING

Examined the water storage capacity of soils under improved organic management. However, in case of waterlogging, organic matter plays also an important role. The bioturbating activity of the macrofauna leaves various so-called conducting macropores in the soil, which are responsible for the drainage of water to deeper soil layers.

Chan *et al.* found a significant reduction in waterlogging after three years under no tillage compared with conventional tillage. The reduction was related to higher density of conducting macropores (140/m^2 and 5/m^2 for no tillage and conventional tillage, respectively), which was associated with higher population density of earthworms (240/m^2 and 36/m^2 for no tillage and conventional tillage, respectively). Plate 24 provide a demonstrationof this effect.

INCREASED YIELDS

Agronomic practices that influence nutrient cycling, especially mineralization and immobilization, result in an immediate productivity gain or loss, which is reflected in the economics of the agricultural system.

Crop yields in systems with high soil organic matter content are less variable than those in soil that are low in organic matter. This is because of the stabilizing effects of favourable conditions of soil properties and microclimate. Improvements in crop growth and vigour stem from direct and indirect effects. Direct effects stem from improvements in nutrient and water content, as described above. Indirect effects stem from a favourable rooting environment and possible weed suppression and a reduction in pests and diseases.

Fig. 6.10 Plate Under No-tillage Conditions, the Internal Pore System of the Soil is Not Distroyed Through Land Preparation Activities and Able to Drain Rainwater from the Surface to Deeper Layers. On Bare Soil the Impact of Raindrops Results in Sealing of the Surface Pores and Thus Poor Drainage.

There are several ways to calculate the rentability of a farm. In general, key parameters such as expenses, income, yield, cost of production and gross margin are used to analyse how well farmers have managed to reach an income. However, in areas where water is a limiting factor, it may be more useful to analyse the conversion of water into yield or even money.

The following example illustrates different ways to compare outcomes of farms and their cropping systems. It is based on the cropping results of 16 farms with a cropped surface of 60 000 ha in Australia. One-third of this surface is cropped in a conventional way (CT), using intensive tillage practices, fire and several passages with herbicides for weed control.

Two-thirds of the area is cropped with the aim to retain as much ground cover as possible, and thus improve the organic matter content of the soil, using specialized planters and herbicides to control weeds, also known as conservation farming (MT). An analysis of the cropping system data for three years yielded the following gross margins for three crops in rotation (wheat, chickpea and sorghum):

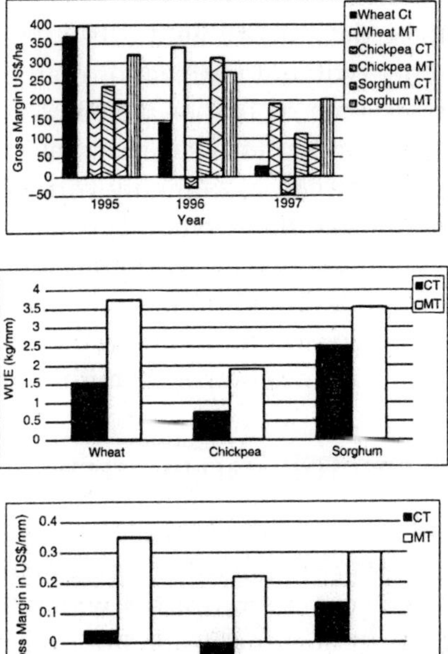

Fig. 6.11

However, when using water-use efficiency (WUE) as means of comparing outputs across farming systems, the results are even more drastic. In this case, rather than the common method of determining WUE by breaking the season into fallow and in-crop components, a total water-use efficiency factor was used. Both grain yield and gross margin are divided by total water to obtain an insight into how well farmers managed the conversion of water into yield and money in the year 1997.

Immobilization of N may occur in systems with crop residue management, especially where C:N ratios of the residues are high (tough, woody materials). This can cause a decrease in maize yield. Figure 6.11 shows the effect of tillage and the preservation of crop residues on maize yield. The preservation of wheat and horse-radish residues on the soil surface led to immobilization of N, which was overcome through the application of N fertilizer.

Based on these data, maize with oats, lupine and vetch as a winter cover crop (without fertilization) can produce a yield that is comparable with or higher than those obtained with conventional tillage and a fertilizer treatment of 90 kg/ha. In this case, the yield increase was highly correlated with the P content of the leaves and the P availability in the soil.

This occurred because of the higher moisture content in the soil under the mulch layer, which led to a higher P uptake by plant roots. Grain crops can also have residual effects on each other through the decomposition of chemical compounds in the residues.

REDUCED HERBICIDE AND PESTICIDE USE

Some people are concerned that intensified systems with reduced or zero tillage will increase herbicide use and in turn lead to increased contamination of water by herbicides. According to Fawcett, total herbicide use in the United States of America declined during the period of adoption of no-tillage systems. He concludes that herbicides are important, but that farmers using conventional tillage methods use similar amounts of herbicides to no-tillage farmers.

In Honduras, a strong decline in the use of herbicides has been observed (Figure 6.12). Farmers who no longer burn their fields prior to preparation spend less money on herbicides. Farmers who have adopted the Quezungual system spend less on herbicides and make savings, both in terms of land preparation and total costs. It is becoming evident that the need for herbicide use diminishes over time in wellmanaged no-tillage cropping systems.

The principal reason is that the system reduces the existing seed bank in the soil by the synergy of two activities: reduction of the production of new seeds through avoidance of flowering and fruit setting; and reduction of seeds that are brought to the surface by tillage practices.

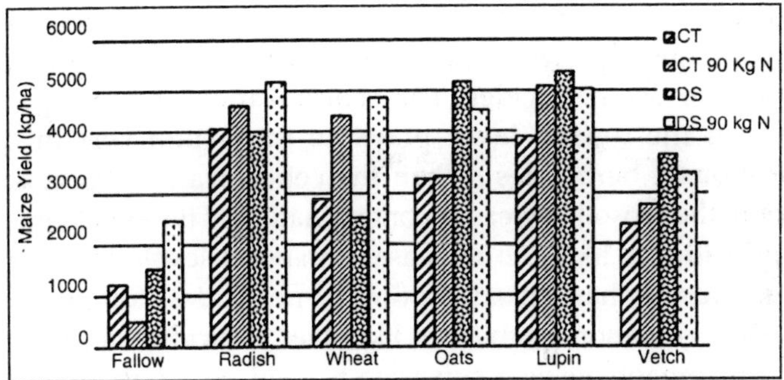

Fig. 6.12 Maize Yield Under Conventional Tillage and Direct Sowing, With and Without 90 kg of N Fertilizer

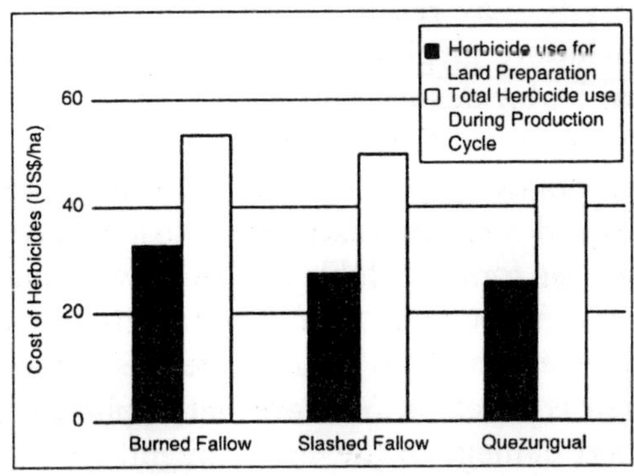

Fig. 6.13 Herbicide Costs in Different Production Systems in Lempira Sur, Honduras

With direct seeding, the reservoir of seeds differs from conventional tillage because:

- The weed seeds remain on the soil surface, where they are susceptible to attacks from insects, birds and soil organisms and to atmospheric influences;
- The soil remains covered by residues, which prevents light reaching the seeds and thus reduces germination;
- Weed seeds already at certain depths are not brought to the surface again, where they could germinate;
- Perennial weeds are no longer redistributed by equipment.

The result is that the soil weed seed store diminishes in time and, consequently, the weed problem also diminishes, as does the need to use herbicides. The concentration of soil organic matter in the topsoil layer plays an important role in the absorption of herbicides. When the concentration of organic matter in the topsoil decreases, contamination of the environment by herbicides is likely to increase. Enhanced levels of organic matter cause enhanced adsorption of pesticides, followed by gradual degradation. Herbicides, like other pesticides, can be used by micro-organisms as a substrate to feed on. Herbicides are broken down in soil and water by micro-organisms into natural acids, NH_4^-, amino acids, carbohydrates, phosphate and CO_2. As microbes cause more rapid degradation of pesticides, enhancing microbial activity may reduce leaching of pesticides. Many herbicides, including glyphosate and paraquat, which are the most common herbicides used in reduced-tillage systems, are bound tightly to clay and organic matter by electrostatic forces and hydrogen bonding. Once they are bound to soil organic matter, the herbicides become inactive and no longer affect plants. Moreover, they can form insoluble complexes with metals in the soils. This also contributes to their rare stability in the soil and low potential for leaching into groundwaters. Some studies have shown that there is no reason to believe that glyphosate may cause any unexpected damage to the environment. However, other studies illustrate negative effects on soil life or its functions. Farming systems that increase soil organic matter content reduce the probability of environmental contamination by herbicides.

INCREASED BIODIVERSITY

Conventional agriculture tends to reduce aboveground and belowground diversity. Thus, it brings about significant changes in the vegetation structure, cover and landscape. The change in vegetal cover during the conversion of forest and pastures to cropping affects plants, animals and micro-organisms. Through increasing specialization of certain plant species and livestock species, some functions may be affected severely, *e.g.* nutrient cycling and biological control. Some non-harvested or associated species profit from the change and become pests. However, many organisms either disappear completely or their numbers are reduced drastically, *e.g.* pollinators and beneficial predators, unless efforts are made to retain a suitable habitat. Associated species can be managed to a certain extent. Through appropriate crop rotations, crop-livestock interactions and the conservation of soil cover, a habitat can be created for a number of species that feed on pests.

This will in turn attract more insects, birds and other animals. Thus, rotations and associations of crops and cover crops as well as hedgerows and field borders promote biodiversity and ecological functions. Because of the complexity and richness of soil biodiversity, the effects of crop and pasture management are less well understood. However, the effects on certain functional groups and, hence, specific soil functions are being recognized increasingly as vital for agricultural productivity and system sustainability. Effect of different tillage practices on scarab beetle-grub holes and their volumes Before the transformation of native grasslands and forests into agricultural areas, a large number of species of scarab beetles and their larvae (white grubs) inhabitated the soils in southern Brazil.

With the transformation of these areas, some of these beetles disappeared, while others became so well adapted and, lacking biological control agents, became important soil-dwelling pests. However, there are other species that can be considered as facultative pests. The larvae prefer feeding on surface litter or surface deposited animal excrements, but may become pests when not enough surface litter is present for them to feed on, like the genera *Diloboderus* and *Bothynus*.

The large beetles create large, permanent galleries (holes) in the soil, down to a depth of more than 1 m, in which they spend most of their lives. The holes may serve as preferential pathways for water infiltration and root growth, and the chambers become niches of increased soil fertility. Both chambers and galleries provide temporary and permanent refuge for many other soil-dwelling invertebrates and microfauna. Research revealed that beetle grub holes were more abundant under no tillage (NT) (8.8-9.6 holes/ m^2) than conventional tillage (CT) (0.7-1.3 holes/m^2). The largest and deepest holes were also found in NT (up to 33.5 mm in diameter and 117 cm deep). Consequently, the total volume of pores opened in NT (450-503 cm^3) was up to almost 10 times greater than in CT (53-107 cm^3). Soil has the ability to restore its life-support processes provided that the disturbance is not too drastic and that sufficient time is allowed for such recovery. Organic matter and biodiversity of soil organisms are the driving factors in this restoration. Decreases in numbers and types of soil organisms and available substrate (organic matter) lead to a decrease in resilience, which in turn can result in a downward spiral of degradation.

RESILIENCE

Resilience can be defined as the ability of a system to recover after disturbance. Soil resilience depends on a balance between restorative and degrading processes. Factors affecting resilience can be grouped in two categories: endogenous and exogenous. Endogenous factors are related to inherent soil properties (rooting depth, texture, structure, topography and drainage) and microclimate and mesoclimate. Exogenous factors include land use and farming system, technological innovations and input management (Lal, 1994). Hence, appropriate agricultural practices can influence these factors in order to enhance soil resilience.

Organic matter and soil organisms play important roles in conserving and improving soil properties that are related to soil resilience. In addition to creating more pores through biological activity, organic matter plays an important role in the formation and stabilization of soil aggregates through bonds between the organic matter and the mineral soil particles.

Soil aggregation can take place through two binding agents:

1. Waste products of bacteria-polysaccharides;
2. Fungal and bacterial hyphae.

The preservation of aggregate stability is important in order to reduce surface sealing and increase water infiltration rates. With increased stability, surface run-off is reduced.

CONSERVATION AGRICULTURE IN ORGANIC MATTER

PRINCIPLES OF CONSERVATION AGRICULTURE

Conservation agriculture makes use of soil biological activity and cropping systems to reduce the excessive disturbance of the soil and to maintain the crop residues on the soil surface in order to minimize damage to the environment and provide organic matter and nutrients.

It is based on four principles:

1. Minimal mechanical soil disturbance, mainly through direct seeding;
2. Permanent soil cover, organic matter supply through the preservation of crop residues and cover crops;
3. Crop rotation for biocontrol and efficient use of the soil profile;
4. Minimal soil compaction.

Although the principles are not new (except for that of minimal disturbance to the soil), it is the fact that they are applied together in conservation agriculture that generates positive outcomes. All the practices (minimal tillage, soil cover and crop rotation) are combined for synergy and added value. In the past, farmers may have tried but abandoned the use of cover crops or zero tillage because of weed problems or yield declines.

There is also a need for improved weed control and rotations for biocontrol of pests and diseases and nutrient uptake. Integration of the conservation agriculture principles provides a win-win situation for both people and the environment, which has catalyzed successful expansion of the area under conservation agriculture worldwide.

Conservation agriculture aims to:

- Provide and maintain optimal conditions in the rootzone (maximum possible depth for crop roots) in order to enable them to grow and function effectively and without hindrance in capturing plant nutrients and water;

- Ensure that water enters the soil so that:
  - Plants have sufficient water to express their potential growth; and
  - Excess water passes through soil to groundwater and streamflow, not over the surface as run-off where it can cause erosion. There is greater potential for increased cropping efficiency as more water is held in the soil profile than under conventional systems;
- Increase beneficial biological activity in the soil in order to:
  - Maintain and rebuild soil architecture for enhanced water entry and distribution within the soil profile;
  - Compete with potential soil pathogens;
  - Contribute to decomposition of organic materials to soil organic matter and various grades of humus; and
  - Contribute to the capture, retention and gradual release of plant nutrients;
- Avoid physical or chemical damage to roots and soil organisms that would disrupt their effective functioning.

ORGANIC MATTER DEPOSITION

The reduction of soil disturbance through zero-tillage, the use of cover crops and the preservation of crop residues on the soil surface result in increased activity of the soil and in the accumulation of organic matter, mainly in the topsoil (Figure 6.14).

An argument often heard in the discussion on conservation agriculture is that it is only feasible in the humid and subhumid tropics and that the generation of sufficient biomass in semi-arid regions is the limiting factor to start implementing conservation agriculture.

However, recent research has shown that even in semi-arid areas of Morocco the application of the principles of conservation agriculture bears its fruits. Mrabet reports higher yields through better water use and improved soil quality; the latter caused by an increase in soil organic C and N and a slight pH decline in the seedzone.

INCREASED CARBON SEQUESTRATION

World soils are important reservoirs of active C and play a major role in the global carbon cycle. As such, soil can be either a source or sink for atmospheric CO_2 depending on land use and the management of soil and vegetation (Figure 6.15).

The conversion of native ecosystems (*e.g.* forests, grasslands and wetlands) to agricultural uses, and the continuous harvesting of plant materials, has led to significant losses of plant biomass and C, thereby increasing the CO_2 level in the atmosphere. In particular, the practice of burning agricultural fields before cultivation has a disastrous effect on soil organic carbon content. Figure 6.14 shows the reduction in soil organic carbon in agricultural fields after 100 years of burning crop residues and weeds compared with an area that was not burned or ploughed during the same period.

The topsoil layer (0-5 cm) represented the greatest carbon loss (36 per cent) compared with the area that was not burned. Soil N stock in the same layer was reduced by 16 per cent. The carbon stock was reduced not only through burning, but because of the whole land-use management, especially a drastic reduction in diversity of species as monocropping was practised.

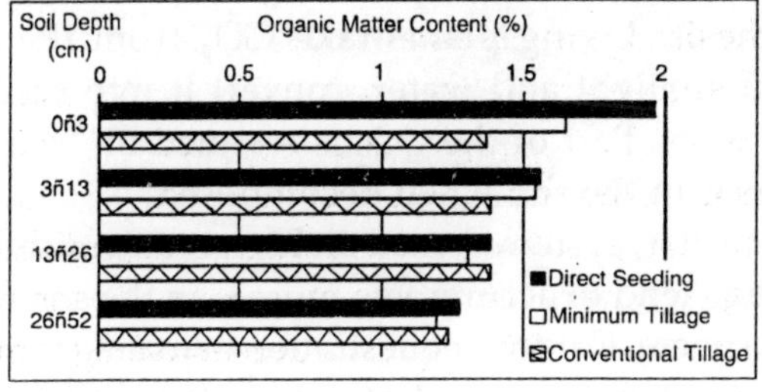

Fig. 6.14 Organic Matter Content of a Soil Under Different Tillage Management

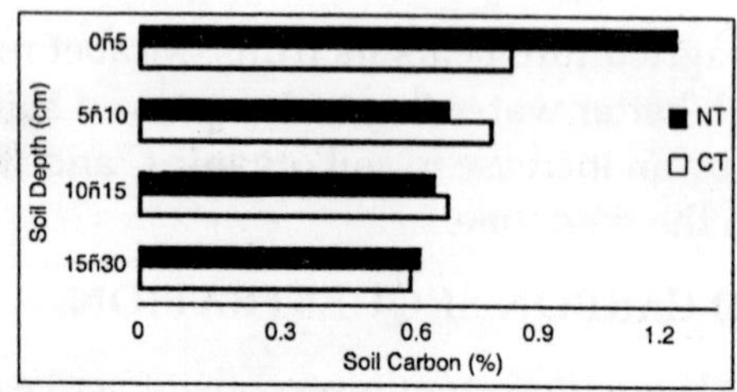

Fig. 6.15 Soil Carbon Concentration at Various Soil Depths Affected by Management System

Note:Conventional(CT)andconservation(NT)agriculture, after two complete cropping cycles (4 years).

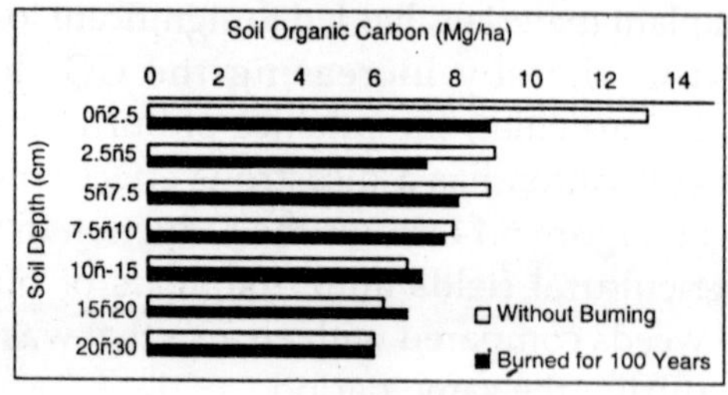

Fig. 6.16 Effect of Crop Residue Burning Once Every Two Years on Soil Carbon Stock

Table lists general practices that determine whether soil will be a sink or a source of atmospheric CO_2. As shown in Table, soil can play a part in mitigating CO_2 levels. This removal process is achieved naturally, and quite effectively, through photosynthesis. Living plants take CO_2 from the air in the presence of sunlight and water, convert it into seeds, leaves, stems and roots. Part of the CO_2 is retained or "sequestered", or stored as C in the soil when decomposed.

In particular, systems based on high crop-residue addition and no tillage tend to accumulate more C in the soil than is lost to the atmosphere. Carbon sequestration in managed soils occurs when there is a net removal of atmospheric CO_2 because C inputs (crop residues, litter, etc.) exceed C outputs (harvested materials, soil respiration, C emissions from fuel and the manufacture of fertilizers, etc.). Management practices that increase soil C comply with a number of principles of sustainable agriculture: reduced tillage, erosion control, diversified cropping system, balanced fertilization, etc.

In the early years of no-tillage systems, the organic matter content of the soil is increased through the decomposition of roots and the contribution of vegetative residues on the surface. This organic material decomposes slowly, and thus the liberation of C to the atmosphere also occurs slowly. In the total balance, net fixation or sequestration of C takes place; the soil is a net sink of C.

Table. Land Use and Land Management Determining Whether Soil Will be a Sink or Source of Atmospheric CO_2

Soil as a source of CO_2	Soil as a sink of CO_2
Soil properties: coarse textured soil, excessive drainage, high susceptibility to erosion	Soil properties: clayey soil, poorly drained ecosystems, depositional sites, including footslopes
Land use: seasonal crops, simple ecosystem, shallow roots and low root-shoot ratio	Land use: perrenial crops, diverse ecosystem, deep roots and high root-shoot ratio
Soil management: intensive tillage based on plough, negative nutrient balance, residue removal and/or burning, continuous cropping, loss of soil and water by runoff and erosion	Soil management: no tillage, positive nutrient balance, mulch farming cover crops in rotation, cycle, soil and water conservation

Figure 6.16 illustrates the fact that some cropping systems can act as a sink for CO_2. In this example, the carbon stock in soils under natural vegetation is used as a reference (steady state: DC = 0). In eight years, the fallow/maize system liberated 4.3 tonnes of CO_2 per hectare. The maize/mucuna system showed a positive balance of almost 20 tonnes of CO_2 per hectare compared with fallow/maize.

Compared with soils under natural vegetation, this means a capture of atmospheric CO_2 of more than 15 tonnes/ha in eight years. Lovato found an increase of 2 tonnes/ha/year over 13 years in a rotation system of oats-common vetch/maiz-cowpea. These figures confirm the potential of conservation agriculture for carbon sequestration. However, the "simple" change from soil tillage to zero tillage is not enough. According to Lovato *et al.* a minimum addition of 4.2 tonnes/ha/year of carbon in vegetative residues in cropping systems and 4.5 tonnes/ha/year in mixed systems of pastures and crops is necessary for maintaining soil organic matter at stable levels.

Even more C can be stored by adding leguminous cover crops to the rotation cycle. This is shown in Figure 6.17, based on two long-term experiments in Rio Grande do Sul, Brazil. Besides addition of C to the soil, legumes add a substantial quantity of N to the soil, which results in increased biomass production of succeeding crops.

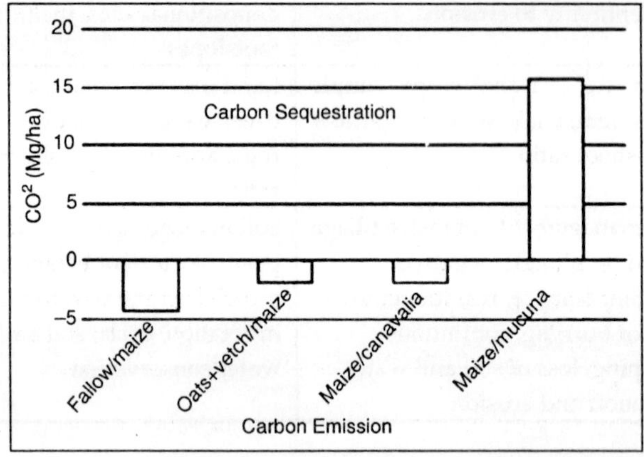

Fig. 6.17 Estimation of Emission and Sequestration of CO_2 (Total Over 8 Years) Under Different Maize Production Systems With Cover Crops Under Direct Sowing Compared With Natural Vegetation in Southern Brazil

Assuming an average carbon accumulation of 0.5 tonne/ha/year, an area like southern Brazil under conservation agriculture would have the potential to sequester 5 million tonnes of C annually, which corresponds to 18 million tonnes of atmospheric CO_2. To compare, Brazil as emitted an estimated 84 million tonnes of CO_2 in 2000.

Recent studies have shown that soil temperature is one of the main climate factors that influence CO_2 emission. High soil temperatures accelerate soil respiration and thus increase CO_2 emission. This has implications for the landscape and land use in a certain area. On convex slopes and hilltops, emission is greater than in foothills, where temperatures are normally lower. The foregoing indicates that not only for soil and water conservation is it important to protect the soil with vegetation (reduction in soil temperature), but that it is advisable to cover the soil also with a view to reducing greenhouse gas emissions.

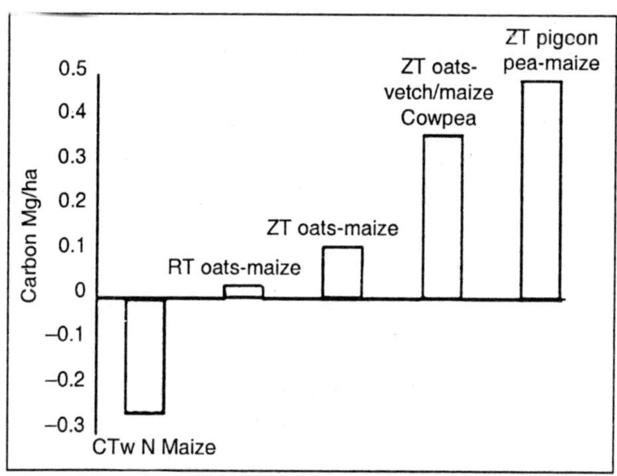

Fig. 6.18 Potential For Carbon Sequestration in Long-Term Experiment in Southern Brazil

Notes: CT = conventional tillage, RT = reduced tillage, ZT = zero tillage, CTwN = conventional tillage without N fertilizer. CT oats-maize (=0).

Chapter 7

Organic form of Soil

NITROGEN

Over 90% of the nitrogen N in the surface layer of most soils occurs in organic forms, with most of the remainder being present as NH_4^- whichis held within the lattice structures of clayminerals. The surface layerof most cultivated soils contains between0.06 and 0.3% N. Peat soils have high N contents to 3.5%.Plant remains and other debris contribute nitrogen N in the form of:

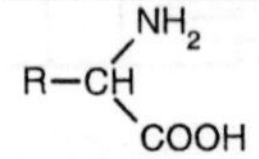

$$R-CH(NH_2)COOH$$

The General Formula of Amino Acids.

(Stevenson 1982)

Fig. 7.1 Amino Acids

Amino acids exist in soil in several different forms, like:

- As free amino acids
  - In the soil solution
  - In soil micropores
- As amino acids, peptides or proteins bound to clay minerals
  - On external surfaces
  - On internal surfaces

- As amino acids, peptides or proteins bound to humic colloids
  - H-bonding and van der Waals' forces
  - In covalent linkage as quinoid-amino acid complexes
- As mucoproteins
- As a muramic acid

Amino acids, being readily decomposed by microorganisms, have only anephemeral existence in soil. Thus the amounts present in the soil solutionat any one time represent a balance between synthesis and destruction by microorganisms.

The free amino acids content of the soil is strongly influenced by weatherconditions, moisture status of the soil, type of plant and stage of growth,additions of organic residues, and cultural conditions.

AMINO SUGARS

Amino sugars occur as structal components of a broad group of substances, the mucopolysaccharides and they have been found in combination with mucopeptides and mucoproteins. Some of the amino sugar material in soil may exist in the form of an alkali-insoluble polysaccharide referred to as chitin. Generally the amino sugars in soil are of microbial origin.From 5 to 10%of the N in the surface layer of most soils can be accounted for in N-containing carbohydrates or amino sugars.

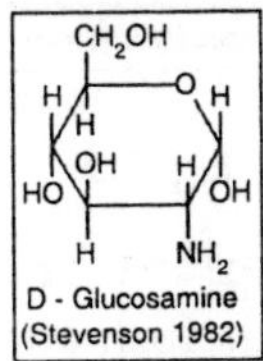

D - Glucosamine
(Stevenson 1982)

Fig. 7.2

NUCLEIC ACIDS

Nucleic acids, which occur in the cells of all living organisms, consist of individual mononucleotide units (base-sugar-phosphate) joined by a phosphoricacid ester linkage through the sugar.Two types: ribonucleic acid (RNA) anddeoxyribonucleic acid (DNA).They have pentose sugar (ribose or deoxyribose),the purine: adenine, guanine and the pyrimidine: cytosine, thymine.RNA contains also the uracil.

The N in purine and pyrimidine bases is usually considered to account forless than 1% of the total soil N. Small amounts of N are extrcted from soil in the form of glycerophosphatides, amines, vitamins, pesticide and pesticide degradation products.

NITROGEN TRANSFORMATION

A key feature of the internal cycle is the biological turnover of N betweenmineral and organic forms through the opposing processes of mineralizationand immobilization. The latter leads to incorporation of N into microbial tissues. Whereas much of this newly immobilized N is recycled through mineralization, some is converted to stable humus forms. The overall reaction leading to incorporation of inorganic forms of N intostable humus forms is depicted on the picture.

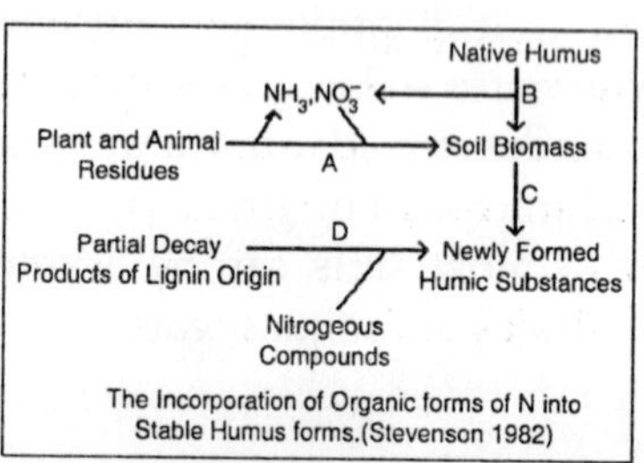

The Incorporation of Organic forms of N into Stable Humus forms.(Stevenson 1982)

Fig. 7.3

Thus the decay of plant and animal residues by microorganisms results in theformation of mineral forms of N (NH_4^+ and NO_3^-) and assimilation of part ofthe C into microbial tissue (reaction A). Part of the native humus undergoes a similar fate (reaction B). Subsequent turnover through mineralization-immobilization leads to incorporation of N into stable humus forms (reaction C). Stabilization of N may also occur through the reaction of partial decay products of lignin with nitrogenous constituents (raection D). Except under unusual circumstances, both mineralization and immobilizationalways function in soil, but in opposite direction.

CHEMICAL REACTION OF AMMONIA AND NITRITE WITH ORGANIC MATTER

The fate of mineral forms of N in soil is determined to some extent bynonbiological reactions involving NH_4^+, $NH_3$and NO_2^- as depicted in fig.

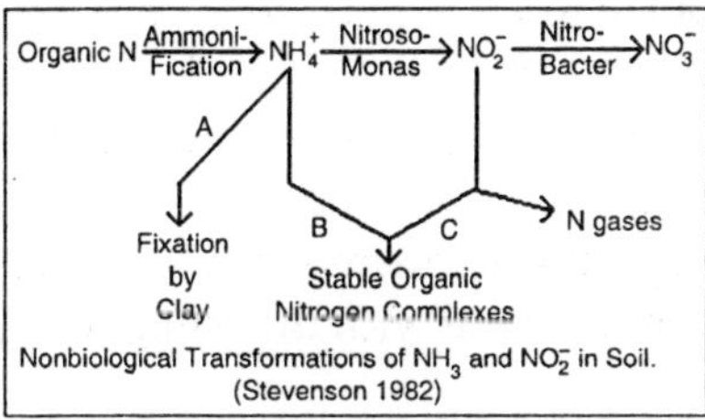

Nonbiological Transformations of NH_3 and NO_2^- in Soil.
(Stevenson 1982)

Fig. 7.4

In addition to NH_4^+ fixation by clay minerals (reaction A), NH_3 and NO_2^- react chemically with organic matter to form stable organic N complexes(reaction B and C). The chemical interaction of NO_2^- with organic matter may lead to the generation of N gases. Although both types of reactions can proceed over a wide pH range, fixation of NH_3^+ is favoured by a high pH (>7.0). In contrast, NO_2^- -organic matter interactions occur most readily under highly acidic conditions (pH of 5.0 to 5.5 or below).

STABILITY OF SOIL ORGANIC NITROGEN

- Proteinaceous constituents are stabilized through their reaction withother organic constituents, such as lignins, tanins, quinons.
- Biologically resistant complexes are formed in soil by chemical reactionsinvolving NH_3^+ or NO_2^- with lignins or humic substances.
- Adsorption of organic N compounds by clay minerals (paricular montmorillinitic types) protects the molecule from decomposition.

- Complexes formed between organic N compounds and polyvalent cations, such as Fe, are biologically stable.
- Some of the organic N occurs in small pores or voids and is physicallyinaccessible to microorganisms.

C/N RATIO

For surface soils, and for the top layer of lake and marine sediments, the ratio generally falls within well-defined limits, usually from about10 to 12. In most soils, the C/N ratio decreases with increasing depth, often attaining values less than 5.0.Native humus would be expected to have a lower C/N ratio than most undecayedplant residues for following reasons. The decay of organic residues by soilorganisms leads to incorporation of part of the C into microbial tissue with the remainder being liberated as CO_2.

As a general rule, about one-third of the applied C in fresh residues will remain in the soil after the first few months of decomposition. The decay process is accompanied by conversion of organic form of N to NH_3 and NO_3^- and soil microorganisms utilize partof this N for synthesis of new cells.

The gradual transformation of plantraw material into stable organic matter (humus) leads to the establishmentof reasonably consistent relationship between C and N. Other factors which may be involved in narrowing of the C/N ratio include chemical fixation of NH_3 or amines by ligninlike substances.

The C/N ratio of virigin soils formed under grass vegetation is normally lower than for soils formed under forest vegetation, and for the latter,the C/N ratio of the humus layers is usually higher than for the mineral soil proper.

Also the C/N ratio of a well-decomposed muck soil is lower than for a fibrous peat. As a general rule it can be said that conditions which encourage decompositionof organic matter result in narrowing of the C/N ratio. The ratio nearly alwaysnarrows sharply with depth in the profile; for certain subsurface soils C/Nratios lower than 5 are not uncommon.

ORGANIC PHOSPHORUS

Phosphorus rank in importance with N and K as major plant nutrients.

Phosphorus compounds in soil can be placed into the following three classes:

1. organic compound of the soil humus,
2. inorganic compounds in which the P is combined with Ca, Mg, Fe, Al and with clay minerals,
3. organic and inorganic P compounds associated with the cells of living matter. Microorganisms are involved in transformations of phosphorus between organic and mineral forms.

From 15 to 80% of the phosphorus in soils occurs in organic forms, the exact amount being dependent upon the nature of the soil and its composition. The higher percentages are typical of peats and uncultivated forest soils. From the standpoint of plant nutrion, phosphorus is adsorbed by plants largely as the negatively charged primary and secondary orthophosphate ions ($H_2PO_4^-$ and HPO_4^{2-}) which are present in the soil solution. Small quantities of soluble organic P compounds are also present in water extracts of soil.

THE PHOSPHORUS CYCLE

In a broad sense, the phosphorus cycle in soil involves the uptake of phosphorus by plants and its return ti the soil in plant and animal residues.

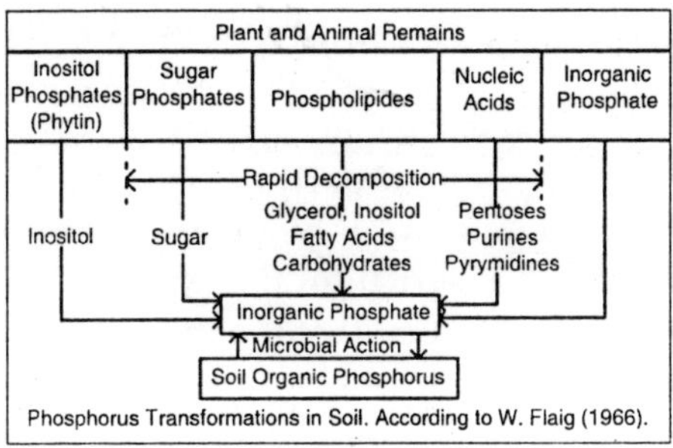

Fig. 7.5

As can be seen from picture three general types of compounds make up the bulk of the organic phosphorus in plants, namely: phytin, phospholipids, and nucleic acids. Approximate recoveries of organic phosphorus in these forms are as follows:

Inositol phosphates	2-50%
Phospholipids	1-5%
Nucleic acids	0.2-0.5%
Phosphoproteins	trace
Metabolic phosphates tace	

When crop residues are returned to the soil, net immobilization of P will occur when the C/organic- P ratio is 300 or mòre; net mineralization will result when the ratio is 200 or less.

FUNCTION OF ORGANIC MATTER IN SOIL

Organic matter contributes to plant growth through its effect on the physical, chemical, and biological properties of the soil.

It has a:

- *Nutritional* function in that it serves as a source of N, P for plant growth
- *Biological* function in that it profoundly affects the activities of microflora and microfaunal organisms
- *Physical* and *physico-chemical* function in that it promotes good soil structure, thereby improving tilth, aeration and retention of moisture and increasing buffering and exchange capacity of soils.

Humus also plays an indirect role in soil through its effect on the uptake of micronutrients by plants and the performance of herbicides and other agricultural chemicals.

It should be emphasized that the importance of any given factor will vary from one soil to another and will depend upon such environmental conditions as climate and crpping history.

AVAILABILITY OF NUTRIENTS FOR PLANTH GROWTH

Organic matter has both a direct and indirect effect on the availability of nutrients for plant growth. In addition to serving as a source of N, P, S through its mineralization by soil microorganisms, organic matter influences the supply of nutrients from other sources (for example, organic matter is required as an energy source for N-fixing bacteria).

A factor that needs to be taken into consideration in evaluating humus as a source of nutrient is the cropping history. When soils are first placed under cultivation, the humus content generally declines over a period of 10 to 30 years until a new equilibrum level is attained. At equilibrium, any nutrients liberated by microbial activity must be compensated for by incorporation of equal amounts into newly formed humus.

EFFECT ON SOIL PHYSICAL CONDITION, SOIL EROSION AND SOIL BUFFERING AND EXCHANGE CAPACITY

Humus has a profound effect on the structure of many soils. The deterioration of structure that accompanies intensive tillage is usually less severe in soils adequately supplied with humus. When humus is lost, soils tend to become hard, compact and cloddy.

Aeration, water-holding capacity and permeability are all favorably affected by humus. The frequent addition of easily decomposable organic residues leads to the synthesis of complex organic compounds that bind soil particles into structural units called aggregates. These aggregates help to maintain a loose, open, granular condition. Water is the better able to infiltrate and percolate downward through the soil.The roots of plants need a continual supply of O_2 in order to respire and grow. Large pores permit better exchange of gases between soil and atmosphere.

Humus usually increases the ability of the soil to resist erosion. First, it enables the soil to hold more water. Even more important is its effect in promoting soil granulation and thus maintaining large pores through which water can enter and percolate downward. From 20 to 70% of the exchange capacity of many soils is caused by colloidal humic substances. Total acidities of isolated fractions of humus range from 300 to 1400 meq/100g.As far as buffer action is concerned, humus exhibits buffering over a wide pH range.

EFFECT ON SOIL BIOLOGICAL CONDITION

Organic matter srves as a source of energy for both macro- and microfaunal organisms. Numbers of bacteria, actinomycetes and fungi in the soil are related in a general way to humus content. Earthworms and other faunal organisms are strongly affected by the quantity of plant residue material returned to the soil. Organic substances in soil can have a direct physiological effect on plant growth. Some compounds, such as certain phenolic acids, have phytotoxic properties; others, such as the auxins, enhance plant growth. It is widely known that many of the factors influencing the incidense of pathogenic

organisms in soil are directly or indirectly influenced by organic matter. For example, a plentiful supply of organic matter may favour the growth of saprophytic organisms relative to parasitic ones and thereby reduce populations of the latter. Biologically active compounds in soil, such as antibiotics and certain phenolic acids, may enhance the ability of certain plants to resist attack by pathogens.

TYPES OF HUMUS IN SOILS

Humus occurs in soils in many types, differentates in regard to morphology and fractional composition. A type of humus is it a morphological form of naturals accumulation of humic substances in profile or on the surface of soil, conditioned by general direction of soil-forming process and humification of organic matter.

A types of humus in terrestrial enviroment are following:

- Mor
- Moder
- Mull

Mor is a type of humus, which occur largely in coniferous forest soils and the moorlands soils. This humus arise under conditions of low-biological activity in soil. The mineralization of organic matter proceed slowly and create layers, which maintain a structure of vegatable material.Acidophilic fungi and low activeinvertebrates participates in transformations of plant residues.

Under these circumstances forms a litter of large thickness. C/N ratio of mor humus is always more than 20, or even 30-40, whereas pH is acid. Moder is a transitional form of humus between mull and moder, characteristic for sod-podzolic soils, loesses and mountain grassland soils.

The organic horizons with moder humus consist of low-thicknessed litter (2-3 cm), which gradually, without bounds, pass on to humus-accumulative horizons. Moder is a type of medium humified humus. Acidophilic fungi and arthropodan participates in transformations of plant residues. C/N ratio equal 15-25. Produced mineral-organic complexes are labile and weakly bounded with mineral portion of soil.

Mull is a type of humus characteristic for chestnut soils, phaeozems, rendzinas and others soils. This type of humus arise under grass vegetation. Mull is a well humified organic matter, which is produce in very biologically active habitat.

This type of humus is characterized by neutral pH, C/N ratio nearing to 10 and ability to creation stable mineral organic complexes. Mull is a type of humus which occurs in soils under cultivation. According to Kononova, the types of humus are divide as follows: First type of humus is characteristic for podzolic soils, grey brown soils and lateric soils under forest communities. In this humus predominate humic acids, thus humic acid/fulvic acid ratio is below 1. Humic acid indicate small extent of aromatic rings condensation and they are approximate to fulvic acids. Considerable hydrophilic properties of humic acids favour to creation of chelates with polyvalent cations and ability to displacement deep into profile of soil. Considerable mobility of this humus favour process of podsolization. Second type of humus is characteristic for phaeozems, rendzinas, black earths and brown soils. Humic acid/fulvic acid ratio is upper than 1, Extent of aromatic rings condensation is high in humic acids, which cause their hydrophobic properties and inability to creation of chelates. Humic acids are strongly connected with mineral portion of soil in this type of humus.

Third type of humus is characteristic for semidesert soils. In this humus predominate fulvic acids fraction, whereas arise of humic acids is limited. Beyond this, humic acids are largely bounded with mineral portion of soil.

HUMUS CONTENT OF SOIL

Humus content in soils fluctuating in broad range.

On humus content have influence the following factors:

- Amount and quality of humus, which get at soil in given bioecological zone
- Tempo of humification process of organic matter
- Tempo of mineralization of humus, which is contain in soil
- Chemical, physico-chemical and physical soil properties
- Amount and quality of mineral compounds contained in soil

Table. Humus Content in Accumulation Horizons of The Main Soil Units in Poland

Division and order	Type,genera and kind	Humus content %
Calcisols:	Calcarious Rendzinas	3.4
		2.1–6.3
	Jurasic Rendzinas	4.4
		1.5–7.0
Phaeozems	Haplic Phaeozems	2.8
		1.8–4.0
Cambisols:	Brown soils formed from sands	1.5
		0.9–2.2
	Brown soils formed from light	1.8
	and medium loams	1.1–3.0

	Brown soils formed from heavy loam	2.5 1.6–3.7
	Brown soils formed from silt formations	1.7 1.3–1.9
	Brown soils formed from loess and loesslike materials	1.9 1.4–2.6
Luvisols:	Grey brown soils formed from silt formations	1.9 1.4–2.4
	Grey brown soils formed from loess and loesslike materials	1.8 1.0–2.5
	Grey brown soils formed from light loam	1.6 1.0–2.6
Podzols	Podzolic soil formed from sands	1.5 1.1–2.0
Gleysols:	Boggy soils formed from silts	1.6 1.2–2.1
Gleysols:	Black earth formed from sands	2.8 1.2–4.1
	Black earth formed from light and medium loams	2.6 1.2–5.7
	Black earth formed from heavy loams and clays	4.9 2.5–5.6
Fluvisols:	Alluvial soils formed from sands	2.9 1.5–5.2
	Alluvial soils formed from silts	3.5 1.7–5.8
	Alluvial soils formed from clays	4.2 2.4–6.8

Chapter 8

Chemical Equilibrium

INTRODUCTION TO CHEMICAL EQUILIBRIUM

Chemical equilibrium applies to reactions that can occur in both directions. In a reaction such as:

$$CH_4(g) + H_2O(g) \leftrightarrow CO(g) + 3H_2(g)$$

The reaction can happen both ways. So after some of the products are created the products begin to react to form the reactants.

At the beginning of the reaction, the rate that the reactants are changing into the products is higher than the rate that the products are changing into the reactants. Therefore, the net change is a higher number of products.

Even though the reactants are constantly forming products and vice-versa the amount of reactants and products does become steady. When the net change of the products and reactants is zero the reaction has reached equilibrium. The equilibrium is a dynamic equilibrium. The definition for a dynamic equilibrium is when the amount of products and reactants are constant. (They are not equal but constant. Also, both reactions are still occurring.)

Some crystals of iodine, $I_2(s)$, are placed in an open gas jar (an open system) maintained at a constant room temperature. A purple vapour is seen to form above the crystals, the crystals gradually disappear, and eventually so does the purple iodine vapour. The experiment is repeated but this time a lid is placed on the gas jar (a closed system). After a short time, a uniform purple vapour forms and some crystals of iodine remain. At the constant temperature the system remains like this indefinitely.

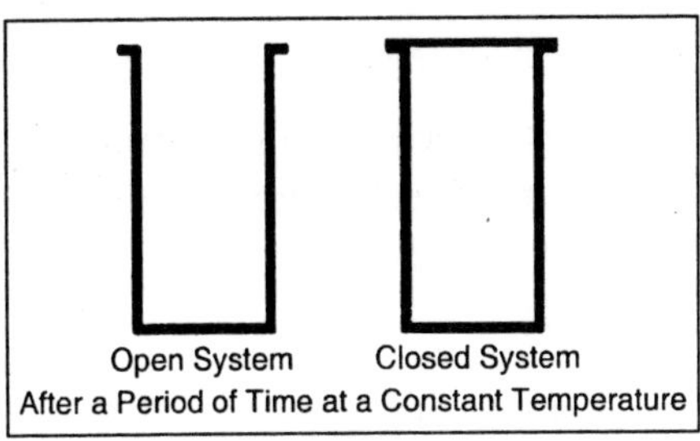

Fig. 8.1

In this closed system an equilibrium has established. The equilibrium is between the iodine solid and the iodine vapour. Here we do not have a static condition but adynamic condition. What is happening all of the time is that I_2 molecules are leaving the solid and entering the vapour phase at the same rate as they are leaving the vapour and entering the solid state. The forward and reverse rates have become the same and the system is at equilibrium. However, this is not a chemical equilibrium but a physical equilibrium.

$$I_2(s) \rightleftharpoons I_2(g)$$

Incidentally, in the closed system above, the pressure of the iodine vapour is equal to its vapour pressure. This does depend on temperature, but is independent of the volume of the vessel and of the amount of iodine solid, provided some solid is always present.

A CHEMICAL REACTION CAN ALSO REACH A STATE OF EQUILIBRIUM

At the start of an experiment you have two sealed flasks maintained at 720K, one containing a mixture of $H_2(g)$ and $I_2(g)$, and the other HI(g) only.

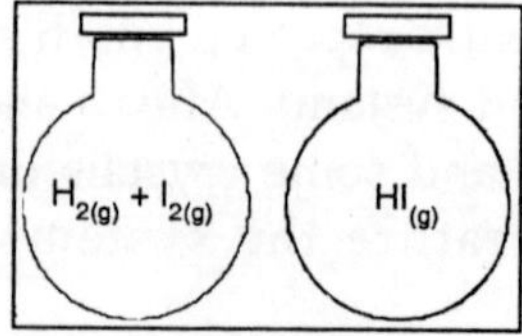

Fig. 8.2

At intervals of time during the experiment the amount of HI(g) in each is measured (and expressed as the per cent of all H and I atoms present as HI).

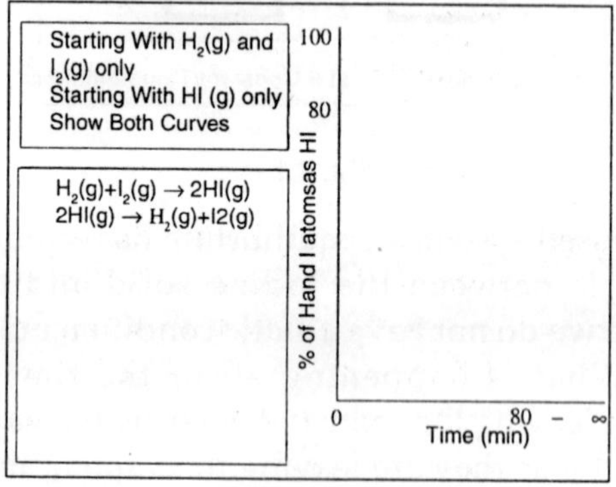

Fig. 8.3

Starting with H_2(g) and I_2(g) only, the two react producing HI(g) in a forward reaction. The rate of this reaction decreases as it proceeds owing to the concentrations of H_2(g) and I_2(g) diminishing. As HI(g) is formed, its molecules begin to react to reform H_2(g) and I_2(g) in a reverse reaction. As the concentration of HI(g) increases as a result of the forward reaction, the rate of the reverse reaction increases. Eventually, the concentrations of all the chemical components will attain values such that the forward and reverse reactions are occuring at the same rate. We now have a reversible reaction that is said to have attained chemical equilibrium. In an equilibrium reaction, all chemical components are both reactants and products, though by tradition those shown on the left of the chemical equation are called reactants and those on the right products.

$$H_2(g) + I_2(g) \rightleftharpoons 2HI(g)$$

Remember, although at equilibrium the forward and reverse reactions may appear to have stopped they are, in fact, taking place all of the time and at the same rate-equilibrium is a dynamic not a static condition.

WHAT QUANTITIES OF REACTANTS AND PRODUCTS ARE PRESENT AT EQUILIBRIUM

Equilibrium refers to equality of forward and reverse reaction rates, not of quantities of reactants and products. In the above chemical equilibrium, 78% of the mixture is in the form of HI(g), so appreciable quantities of both reactants and products are present.

Some reactions establish equilibrium after the formation of only minute amounts of products. In others, only minute amounts of reactants remain at equilibrium. The composition of the equilibrium mixture is sometimes referred to in terms ofequilibrium position.

HOW LONG DOES IT TAKE FOR EQUILIBRIUM TO BE REACHED

In the case of HI(g), the mixture reaches equilibrium in less than 2 hours at 720 K. Other reactions may take only a tiny fraction of a second, and there are many more that would be nowhere close to equilibrium after a million years.

THE EQUILIBRIUM CONSTANT

Now consider we have a mixture of reactants and products that is not yet at equilibrium, or one that is at equilibrium. Which way will the reaction go in order to reach equilibrium? How far will the reaction go, that is, what will be the amounts of reactants and products at equilibrium?

If we make changes to the system, such as to its volume, or add more of a reactant, how will the equilibrium composition be affected? To be able to answer these questions we have to know about the Equilibrium Constant. Now back to the equilibrium involving $H_2(g)$, $I_2(g)$ and HI(g) at 720K.

$$H_2(g) + I_2(g) \rightleftharpoons 2HI(g)$$

In four separate experiments at 720 K, chemical equilibrium was allowed to establish. Two were started with different concentrations of $H_2(g)$ and $I_2(g)$ only in each, and two with differing initial concentrations of HI(g) only. In each of the four experiments, the equilibrium concentration of $H_2(g)$, $I_2(g)$ and HI(g) were measured.

Table. Equilibrium Concentrations (Mol Dm^{-3})

$[H_2(g)]_{eqm}$	$[I_2(g)]_{eqm}$	$[HI(g)]_{eqm}$
1.14×10^{-2}	0.12×10^{-2}	2.52×10^{-2}
0.92×10^{-2}	0.20×10^{-2}	2.96×10^{-2}
0.34×10^{-2}	0.34×10^{-2}	2.35×10^{-2}
0.86×10^{-2}	0.86×10^{-2}	5.86×10^{-2}

If we substitute these equilbrium concentrations into the expression below:

$$\frac{\left[HI(g)\right]^2 eqm}{\left[H_2(g)\right]eqm\left[I_2(g)\right]eqm}$$

then a constant value, within experimental error, is obtained. The square brackets represent concentrations in mol dm^{-3}.

The value of this ratio of equilibrium concentrations (in mol dm^{-3}) is known as the equilibrium constant, represented by the symbol K_c.

$$K_c = \frac{\left[HI(g)\right]^2 eqm}{\left[H_2(g)\right]eqm\left[I_2(g)\right]eqm} (= 46.8 \text{ at } 720K)$$

Many other chemical equilibrium reactions have been studied and in each case an equilibrium constant relating to the stoichiometric chemical equation has been calculated. This observation can be universally applied, in what is sometimes called the *Law of Chemical Equilibrium,* as illustrated by the general chemical equilibrium:

$$aA_{(g)} + bB_{(g)} \rightleftharpoons cC_{(g)} + dD_{(g)}$$

$$K_c = \frac{[C(g)]_{eqm}^{c}[D(g)]_{eqm}^{d}}{[A(g)]_{eqm}^{a}[B(g)]_{eqm}^{b}}$$

CAN THE EQUILBRIUM CONSTANT BE EXPRESSED IN OTHER TERMS

The Ideal Gas Equation shows that the pressure of a gas is proprtional to its concentration.

$$pV = nRT$$

where p is pressure of a particular gas (its partial pressure) in an equilibrium mixture, V is the total volume, n is the number of moles of the particular gas, R is the general gas constant, and T is the absolute temperature.

In the above equation where the temperature is also constant,

$$P \alpha n/V$$

The equilibrium constant can therefore also be expressed in terms of partial pressures, and is denoted as K_p:

$$K_p = \frac{pC(g)_{eqm}^{c} \; pD(g)_{eqm}^{d}}{pA(g)_{eqm}^{a} \; pB(g)_{eqm}^{b}}$$

DOES THE EQUILBRIUM CONSTANT HAVE UNITS

The equilibrium constant (K_c or K_p) may or may not have units; this depends precisely upon the equilibrium expression.

WHAT VALUES CAN THE EQUILIBRIUM CONSTANT HAVE

Equilibrium constants come in all sizes, from almost zero (like 10^{-50}) to extremely large (like 10^{-50}). When K (K_c or K_p) is much greater than 1, the partial pressures or concentrations of the products are large in comparison to those of the reactants.

EQUILIBRIUM CONSTANT DEPEND ON TEMPERATURE

Yes. An equilibrium constant is constant only at a constant

temperature. Changing the temperature of a chemical system at equilibrium will change the value of the equilibrium constant, and also the position of the equilibrium.

ARE K_c AND K_p RELATED

Yes. The relationship depends on the reaction and how it is written. Specifically, it depends on the number of moles of *gaseous* reactants and products.

The equation is:

$$K_p = K_c(RT)^{\Delta n}$$

where Δn is the number of moles of gaseous products minus the number of moles of gaseous reactants, R is the gas constant, and T isthe absolute temperature at which the equilibrium exists.

Now let's make some use of the equilibrium constant...

Return to: The Equilibrium Constant

Which way will the reaction go in order to reach equilibrium?

We have a mixture of SO_2(g), O_2(g), and SO_3(g) at 1000K which has still to reach equilibrium. The partial pressures are p_{SO2} = 0.48 atm, p_{O2} = 0.18 atm, and p_{SO3} = 0.72 atm. K_p = 3.40 atm^{-1}. The reaction is:

$$2SO_2(g) + O_2(g) \rightleftharpoons 2SO_3(g)$$

Which way must the reaction go to reach equiilibrium?

Answer:

Calculate a value (Q) for the reaction which can be compared with the equilibrium constant, K_p. This is given by the expression:

$$Q = \frac{p_{SO_3^2(g)}}{p_{SO_2^2(g)} p_{O_2(g)}}$$

The value of Q calculated is 12.50 atm^{-1}. Since Q is greater than K_p, to reach equilibrium, the equilibrium must go from right to left.

How far will the reaction go, that is, what will be the amounts of reactants and products at equilibrium?

A sample of air is compressed so that $[N_2(g)] = 0.80$ mol dm^{-3} and $[O_2(g)] = 0.20$ mol dm^{-3}. (It might help to imagine the container has a volume of 1 dm^3.) It is then heated to 1500 K and allowed to reach equilibrium.

$$N_2(g) + O_2(g) \rightleftharpoons 2NO(g)$$

How much of each reactant and product will be present at equilibrium? $K_c = 1.0 \times 10^{-5}$ at 1500K.

Answer:

Look at the value of Kc. Because it is such as small value, the amounts of $N_2(g)$ and $O_2(g)$ that have reacted on reaching equilibrium will be negligible compared to the original amounts present. You can assume that their equilibrium concentrations are the same as the initial concentrations. There was no NO(g) at the start, so the amount that forms will be signficant. Equal molar amounts of $N_2(g)$ and $O_2(g)$ react on reaching equilibrium, so let this amount be equal to x. From the balanced chemical equation, the amount of NO(g) at equilibrium will be 2x. Find the value of x by substituting the equilibrium concentration of each chemical component into the equilibrium expression for the reaction:

$$K_c = \frac{[NO(g)]^2_{eqm}}{[N_2(g)]_{eqm}[O_2(g)]_{eqm}} = 1.0 \times 10^{-5} \text{ at } 1500k$$

The value of x calculates to be 6.32×10^{-4}, so the equilibrium concentration of NO(g) is 1.26×10^{-3} mol dm^{-3}.

If we make changes to the system, such as to its volume, or add more of a reactant, how will the equilibrium composition be affected?

It is of great practical importance to be able to predict, and to control, the equilibrium composition of a chemical reaction. In an industrial process it may be desirable to convert as much as possible of reactants to prodcuts. At equilibrium the product yield may not be satisfactory. But, the composition of the equilibrium mixture (equilibrium position) is dependent on certain conditions, such as temperature and pressure, and it might be possible to change these to achieve a better yield. Le

Chatelier's Principle helps us deal with this. Now consider another chemical reaction in a state of equilibrium at a constant temperature.

$$N_2(g) + 3H_2(g) \rightleftharpoons 2NH_3(g)\ \Delta H^\circ_{f,\,298} = -\ 46.0\ kJ\ mol^{-1}$$

In a closed system and at a pressure of 250 atmospheres in the presence of a finely divided iron catalyst at 500 °C, about 15 per cent of the gases are converted into ammonia at equilibrium.

We can impose the following changes [both an increase and decrease] upon this chemical system at equilibrium:

- Amount of a substance/ concentration of a reactant or product.
- Pressure/ Volume (pressure and volume are inversely related).
- Temperature.

In general terms, this is what Le Chatelier's Principle says... If a change is imposed upon a chemical system at equilibrium then it will respond in such a way as to undo, in part, the effect of the change imposed upon it.

Here are Some Examples

Example 1

The reaction is at equilibrium in a closed vessel of fixed volume. The temperature is constant. Suddenly some nitrogen gas is added. The concentration of nitrogen has increased. The partial pressure of nitrogen is raised. The system is no longer at equilibrium. Le Chatelier's Principle says that the equilibrium will adjust so as to reduce the concentration of the nitrogen.

This can only be achieved if the nitrogen reacts. The nitrogen can only react with hydrogen forming ammonia. The equilibrium therefore shifts from left to right. A new equilibrium position is established. Since the system responds 'to undo, in part, the effect of the change', at the new equilibrium position there will be slightly more nitrogen, less hydrogen, and more ammonia present.

Example 2

The reaction is at equilibrium in a closed vessel. The temperature is constant. The volume of the vessel is suddenly reduced. This is equivalent to increasing the total pressure of the system. The system is no longer at equilibrium. Le Chatelier's Principle says that the equilibrium will adjust so as to reduce the pressure. It can only achieve this by shifting in the direction of fewer moles of gas (fewer gaseous molecules).

There are 2 moles of gas on the right of the equilibrium and 4 on the left. The equilibrium therefore shifts from left to right. A new equilibrium position is established. The yield of ammonia is increased.

Example 3

The reaction is at equilibrium in a closed vessel of fixed volume. The temperature of the system is suddenly reduced. The system is no longer at equilibrium. Le Chatelier's Principle says that the equilibrium will adjust so as to bring about a rise in temperature of the system. A reaction needs to occur so as to produce heat for this to happen. The equilibrium must shift in the exothermic direction, that is, from left to right. A new equilibrium position is established. The yield of ammonia is increased.

GETTING THE BEST YIELD OF AMMONIA

It seems that by raising the pressure of the system and at the same time reducing its temperature we can maximise the yield of ammonia at equilibrium. This is not without its problems. Maintaining a higher pressure is expensive; more robust plant (thicker pipes and stronger joints, for example) is required, and more electricity would have to be used to drive pumps to maintain this pressure. Lowering the temperature slows the rate at which chemical reactions take place resulting in a longer wait for equilibrium to be achieved. A 'compromise' has to be settled upon.

DOES A CATALYST AFFECT THE EQUILIBRIUM POSITION

No. The addition or removal of a catalyst does not cause a shift in equilibrium. A catalyst *is a substance that affects the rate of a reaction, but is not consumed in the reaction*. It does this by providing an alternative mechanism for the reaction but of lower activation enthalpy. It therefore changes the rate of approach of equilibrium and so affects the forward and reverse reaction rates in the same way. The composition of the equilibrium mixture is unchanged.

HETEROGENEOUS EQUILIBRIUM

So far we have considered only chemical equilibria involving all gases, that is, homogeneous equilibria. An equilibrium involving more than one phase-gas and solid, for example, or liquid and solid-is said to be heterogeneous. For example:

$$H_2(g) + I_2(s) \rightleftharpoons 2HI(g)$$

In the above example, the equilibrium expressions for this reaction are:

$$K_c = \frac{[HI(g)]^2 \text{ eqm}}{[H_2(g)]\text{eqm}} \quad K_p = \frac{p_{HI(g)^2 eqm}}{p_{H2(g)eqm}}$$

Notice that I_2 is missing from the equilibrium expressions. This is because, at room temperature, iodine is a solid. The composition of the equilibrium mixture of $H_2(g)$ and $HI(g)$ is independent of the amount of solid iodine present, *as long as some of it is always present*. [The partial pressure of $I_2(g)$ (which is its vapour pressure) in equilibrium with $I_2(s)$ is constant at a constant temperature, and is independent of the volume of the container and of the quantity of solid.]

EQUILIBRIUM IN SOLUTION

Here we will consider homogeneous and heterogeneous chemical equilibria in which the solvent is much more abundant than all the other components put together. The solvent may also be a reactant or product itself.

First of all, let's look at an equilibrium that is not in solution:

$$CH_3COOH(I) + CH_3CH_2OH(I) \rightleftharpoons CH_3COOCH_2CH_3(I) + H_2O(I)$$

$$K_c = \frac{[CH_3COOCH_2CH_3(I)]\text{eqm } [H_2O(I)]\text{eqm}}{[CH_3CH_2OH(I)]\text{eqm}' \ [CH_3COOH(I)]\text{eqm}}$$

In this case, water is not a solvent but a product only.

Now consider the same equilibrium in very dilute aqueous solution. The chemical equation is:

$$CH_3COOH(aq) + CH_3CH_2OH(aq) \rightleftharpoons CH_3COOCH_2CH_3(aq) + H_2O(l)$$

And the equilibrium expression is:

$$K_c = \frac{[CH_3COOCH_2CH_3(aq)]eqm}{[CH_3CH_2OH(aq)]eqm\ [CH_3COOH(aq)]eqm}$$

Water is now both a product and a solvent. In reaching a state of equilibrium the concentration of the water changes so little that it is effectively constant. For this reason its value is taken along with the equilibrium constant, K_c.

Here are some more examples of chemical equilibria in aqueous solution:

$$CH_3COOH_{(aq)} + H_2O_{(l)} \rightleftharpoons CH_3COO^-_{(aq)} + H_3O^+_{(aq)}$$

$$K_a = \frac{[H_3O^+(aq)]eqm[CH_3COO^-(aq)]eqm}{[CH_3COOH(aq)]eqm}$$

K_a is called the acid dissociation constant. Again, the concentration of water changes negligibly on reaching equilibrium and so its value is taken along with the equilibrium constant (K_c) to form the 'modified' equilibrium constant, K_a.

$$NH_{3(aq)} + H_2O_{(l)} \rightleftharpoons NH_4^+{}_{(aq)} + OH^-{}_{(aq)}$$

$$K_b = \frac{[NH_4^+(aq)]eqm[OH^-(aq)]eqm}{[NH_3(aq)]eqm}$$

K_b is called the base dissociation constant. Again, the concentration of water changes negligibly on reaching equilibrium and so its value is taken along with the equilibrium constant (K_c) to form the 'modified' equilibrium constant, K_b.

$$AgCl_{(s)} \rightleftharpoons Ag^+{}_{(aq)} + Cl^-{}_{(aq)}$$

$$K_s = [Ag^+_{(aq)}]eqm[Cl^-_{(aq)}]eqm$$

K_s is the Solubility Product. It is the equilibrium constant for the equilibrium that exists between a slightly soluble salt and its ions in saturated solution. The solid, AgCl, is omitted from the equilibrium expression. As long as there is some solid present the concentration of the saturated solution of ions remains constant at a given temperature.

Chemical change is one of the two central concepts of chemical science, the other being *structure*. The very origins of Chemistry itself are rooted in the observations of transformations such as the combustion of wood, the freezing of water, and the winning of metals from their ores that have always been a part of human experience.

It was, after all, the quest for some kind of constancy underlying change that led the Greek thinkers of around 200 BCE to the idea of elements and later to that of the atom.

Fig. 8.4

It would take almost 2000 years for the scientific study of matter to pick up these concepts and incorporate them into what would emerge, in the latter part of the 19th century, as a modern view of chemical change.

CHEMICAL CHANGE: HOW FAR, HOW FAST

Chemical change occurs when the atoms that make up one or more substances rearrange themselves in such a way that new substances are formed. These substances are the *components* of the *chemical reaction system*; those components which decrease in quantity are called *reactants*, while those that increase are *products*.

A given chemical reaction system is defined by a *balanced net chemical equation* which is conventionally written as *reactants → products*

The first thing we need to know about a chemical reaction represented by a balanced equation is whether it can actually take place. If the reactants and products are all substances capable of an independent existence, then in principle, the answer is always "yes". This answer must be qualified, however, by the following considerations:

That is, what fraction of the reactants are converted into products? Some reactions convert essentially 100% of reactants to products, while for others the quantity of products may be undetectable.

Many are somewhere in between, meaning that significant quantities of all components remain at the end. Later on, in another part of the course, you will learn that the tendency of a reaction to occur can be predicted entirely from the properties of the reactants and products through the laws of thermodynamics.

Some reactions are over in microseconds; others take years. The speed of any one reaction can vary over a huge range depending on the temperature, the state of matter (gas,liquid, solid) and the presence of a catalyst.

Unlike the question of completeness, there is no simple way of predicting reaction speed. What happens, at the atomic or molecular level, when reactants are transformed into products? What intermediate species (those that are produced but later consumed so that they do not appear in the net reaction equation) are involved? This is the *microscopic,* or kinetic view of chemical change, and cannot be predicted by theory as it is presently developed and must be inferred from the results of experiments.

A reaction that is thermodynamically possible but for which no reasonably rapid mechanism is available is said to be *kinetically limited.* Conversely, one that occurs rapidly but only to a small extent is *thermodynamically limited.* As you will see later, there are often ways of getting around both kinds of limitations, and their discovery and practical applications constitute an important area of industrial chemistry.

Fig. 8.5 Textile Artist Hilary Rice Offers This View of Equilibrium Which Nicely Evokes its Dual Character of Internal Dynamics Contained Within a Static Exterior.

Basically, the term refers to what we might call a "balance of forces". In the case of mechanical equilibrium, this is its literal definition.

A book sitting on a table top remains at rest because the downward force exerted by the earth's gravity acting on the book's mass (this is what is meant by the "weight" of the book) is exactly balanced by the repulsive force between atoms that prevents two objects from simultaneously occupying the same space, acting in this case between the table surface.

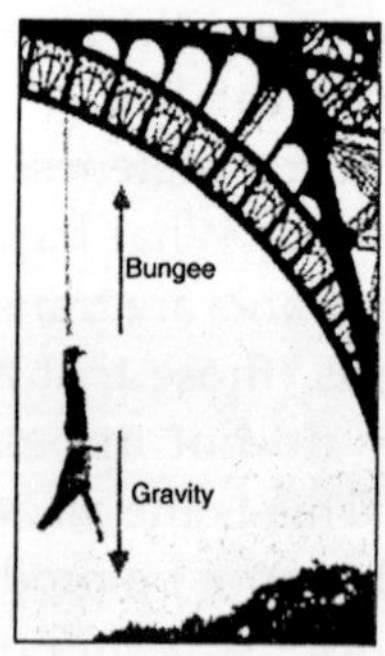

Fig. 8.6

An object is in a state of mechanical equilibrium when it is either static (motionless) or in a state of unchanging motion. From the relation $f = ma$, it is apparent that if the net force on the object is zero, its acceleration must also be zero, so if we can see that an object is not undergoing a change in its motion, we know that it is in mechanical equilibrium.

THERMAL EQUILIBRIUM

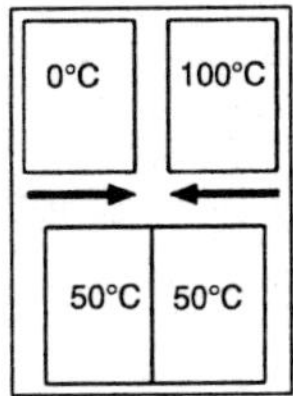

Fig. 8.7

Another kind of equilibrium we all experience is thermal equilibrium. When two objects are brought into contact, heat will flow from the warmer object to the cooler one until their temperatures become identical. Thermal equilibrium arises from the tendency of thermal energy to become as dispersed or "diluted" as possible. A metallic object at room temperature will feel cool to your hand when you first pick it up because the thermal sensors in your skin detect a flow of heat from your hand into the metal, but as the metal approaches the temperature of your hand, this sensation diminishes. The time it takes to achieve thermal equilibrium depends on how readily heat is conducted within and between the objects; thus a wooden object will feel warmer than a metallic object even if both are at room temperature because wood is a relatively poor thermal conductor and will therefore remove heat from your hand more slowly.

Thermal equilibrium is something we often want to avoid, or at least postpone; this is why we insulate buildings, perspire in the summer and wear heavier clothing in the winter.

CHEMICAL EQUILIBRIUM

When a chemical reaction takes place in a container which prevents the entry or escape of any of the substances involved in the reaction, the quantities of these components change as some are consumed and others are formed. Eventually this change will come to an end, after which the composition will remain unchanged as long as the system remains undisturbed. The system is then said to be in its *equilibrium state,* or more simply, "at equilibrium".

Why reactions go towards equilibrium

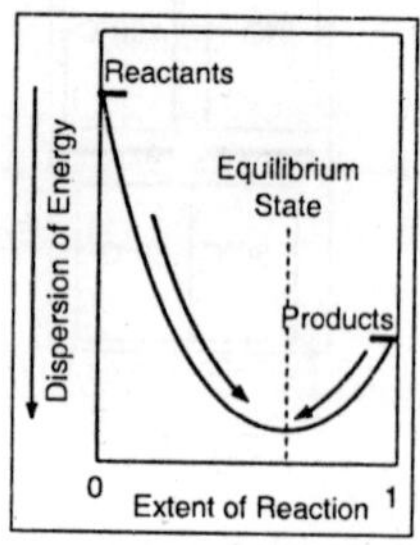

Fig. 8.8

What is the nature of the "balance of forces" that drives a reaction towards chemical equilibrium? It is essentially the balance struck between the tendency of energy to reside within the chemical bonds of stable molecules, and its tendency to become dispersed and diluted. Exothermic reactions are particularly effective in this, because the heat released gets dispersed in the infinitely wider world of the surroundings. In the reaction represented here, this balance point occurs when about 60% of the reactants have been converted to products. Once this equilibrium state has been reached, no further net change will occur. (The only spontaneous changes that are allowed follow the arrows pointing towards maximum dispersal of energy.) A more complete explanation of this must be deferred until the discussion of thermodynamics in a later chapter.

Fig. 8.9 Chemical Equilibrium is Something You Definitely Want to Avoid For Yourself as Long as Possible.

The myriad chemical reactions in living organisms are constantly moving *toward*equilibrium, but are prevented from getting there by input of reactants and removal of products. So rather than being in equilibrium, we try to maintain a "steady-state" condition which physiologists call *homeostasis*—maintenance of a constant internal environment.

For the time being, it's very important that you know this definition: A chemical reaction is in equilibrium when there is no tendency for the quantities of reactants and products to change. The direction in which we write a chemical reaction (and thus which components are considered reactants and which are products) is arbitrary.

Thus the two equations:

1. $H_2 + I_2 \rightarrow 2HI$ "Synthesis of hydrogen iodide"
2. $2\,HI \rightarrow H_2 + I_2$ "Dissociation of hydrogen iodide"

represent the same chemical reaction system in which the roles of the components are reversed, *and both yield the same mixture of components when the change is completed.*

This last point is central to the concept of chemical equilibrium. It makes no difference whether we start with two moles of HI or one mole each of H_2 and I_2; once the reaction has run to completion, the quantities of these two components will be the same. In general, then, we can say that thecomposition of a chemical reaction system will tend to change in a direction that brings it closer to its equilibrium composition. Once this equilibrium composition has been attained, no further change in the quantities of the components will occur as long as the system remains undisturbed.

The figure 8.10 show how the concentrations of the three components of this chemical reaction change with time. Examine the two sets of plots carefully, noting which substances have zero initial concentrations, and are thus "products" of the reaction equations shown. Satisfy yourself that these two sets represent the same *chemical reaction system*, but with the reactions occurring in opposite directions. Most importantly, note how the final (equilibrium) concentrations of the components are the same in the two cases.

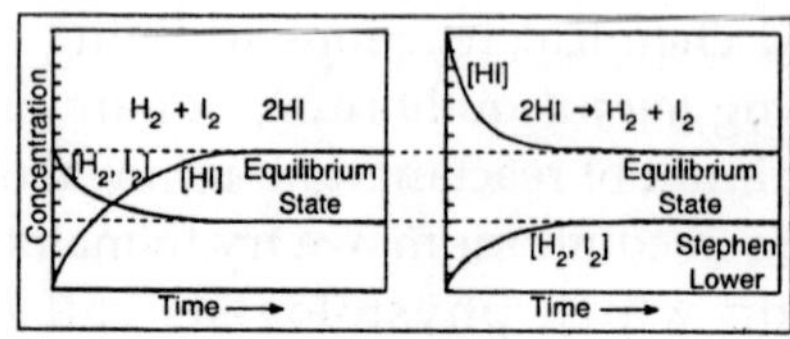

Fig. 8.10

Whether we start with an equimolar mixture of H_2 and I_2 (left) or a pure sample of hydrogen iodide (shown on the right, using twice the initial concentration of HI to keep the number of atoms the same), the composition after equilibrium is attained (shaded regions on the right) will be the same.

The equilibrium composition is independent of the direction from which it is approached.

What is a Reversible Reaction?

A chemical equation of the form A $\rightarrow$ B represents the transformation of A into B, but it does not imply that *all* of the reactants will be converted into products, or that the *reverse* reaction B $\rightarrow$ A cannot also occur.

In general, *both* processes (forward and reverse) can be expected to occur, resulting in an *equilibrium mixture* containing finite amounts of *all* of the components of the reaction system. (We use the word *components* when we do not wish to distinguish between reactants and products.)

If the equilibrium state is one in which significant quantities of both reactants and products are present, then the reaction is said to *incomplete* or *reversible.*

The latter term is preferable because it avoids confusion with "complete" in its other sense of being completed or finished, implying that the reaction has run its course and is now at equilibrium.

- If it is desired to emphasize the reversibility of a reaction, the single arrow in the equation is replaced with a pair of hooked lines pointing in opposite directions, as in A $\rightleftharpoons$ B.
  - Note that there is no fundamental difference

between the meanings of A→B and A $\rightleftharpoons$ B. Some older textbooks just use A = B.

- A reaction is said to be *complete* or *quantitative* when the equilibrium composition contains no significant amount of the reactants. However, a reaction that is complete when written in one direction is said "not to occur" when written in the reverse direction.

In principle, all chemical reactions are reversible, but this reversibility may not be observable if the fraction of products in the equilibrium mixture is very small, or if the reverse reaction is very slow (the chemist's term is *"kinetically inhibited"*) How did Napoleon Bonaparte help discover reversible reactions?

Fig. 8.11

Napoleon recruited the eminent French chemist Claude Louis Berthollet (1748-1822) to accompany him as scientific advisor on the most far-flung of his campaigns, the expedition into Egypt in 1798. Once in Egypt, Berthollet noticed deposits of sodium carbonate around the edges of some the salt lakes found there. He was already familiar with the reaction,

$$Na_2CO_3 + CaCl_2 \rightarrow CaCO_3 + 2NaCl$$

He immediately realised that the Na_2CO_3 must have been formed by the reverse of this process brought about by the very high concentration of salt in the slowly-evaporating waters. This led Berthollet to question the belief of the time that a reaction could only proceed in a single direction. His famous textbook *Essai de statique chimique* (1803) presented his speculations on chemical affinity and his discovery that an excess of the product of a reaction could drive it in the reverse direction.

Unfortunately, Berthollet got a bit carried away by the idea that a reaction could be influenced by the amounts of substances present, and maintained that the same should be true for the compositions of individual compounds. This brought him into conflict with the recently accepted Law of Definite Proportions (that a compound is made up of fixed numbers of its constituent atoms), so his ideas (the good along with the bad) were promptly discredited and remained largely forgotten for 50 years. (Ironically, it is now known that certain classes of compounds do in fact exhibit variable composition of the kind that Berthollet envisioned.)

What is the Law of Mass Action?

Berthollet's ideas about reversible reactions were finally vindicated by experiments carried out by others, most notably the Norwegian chemists (and brothers-in-law) Cato Guldberg and Peter Waage. During the period 1864-1879 they showed that an equilibrium can be approached from either direction, implying that any reaction $aA + bB \rightarrow cC + dD$ is really a competition between a "forward" and a "reverse" reaction. When a reaction is at equilibrium, the rates of these two reactions are identical, so no *net* (macroscopic) change is observed, although individual components are actively being transformed at the microscopic level.

Equilibrium is Dynamic

The "active masses" are essentially the*concentrations* of the reactants and products that combine directly in the manner represented by the reaction equation. The meaning of the terms on the right sides of the equations is simply that the rate is proportional to the concentrations of the components.

Guldberg and Waage showed that for a reaction $aA + bB \rightarrow cC + dD$, the rate (speed) of the reaction in either direction is proportional to what they called the "active masses" of the various components:

$$\text{Rate of forward reaction} = k_f[A]^a[B]^b$$
$$\text{Rate of reverse reaction} = k_r[C]^c[D]^d$$

in which the proportionality constants *k* are called *rate constants* and the quantities in square brackets represent concentrations. If we combine the two reactants A and B, the forward reaction starts immediately; then, as the products C and D begin to build up, the reverse process gets underway. As the reaction proceeds, the rate of the forward reaction diminishes while that of the reverse reaction increases. Eventually the two processes are proceeding at the same rate, and the reaction is at equilibrium:

Rate of forward reaction = rate of reverse reaction

$$k_f[A]^a[B]^b = k_r[C]^c[D]^d$$

It is very important that you understand the significance of this relation. The equilibrium state is one in which there is no net change in the quantities of reactants and products. But do not confuse this with a state of "no change"; at equilibrium, the forward and reverse reactions continue, but *at identical rates*, essentially cancelling each other out. Equilibrium is macroscopically static, but is microscopically dynamic!

To further illustrate the dynamic character of chemical equilibrium, suppose that we now change the composition of the system previously at equilibrium by adding some C or withdrawing some A (thus changing their "active masses"). The reverse rate will temporarily exceed the forward rate and a change in composition ("a shift in the equilibrium") will occur until a new equilibrium composition is achieved. The composition of the equilibrium state depends on the ratio of the forward- and reverse rate constants. Be sure you understand the difference between the *rate* of a reaction and a *rate constant.* The latter, usually designated by *k*, relates the reaction rate to the concentration of one or more of the reaction components—for example, *rate* = *k* [A]. At equilibrium the rates of the forward and reverse processes are identical, but the rate *constants* are generally different. To see how this works, consider the simplified reaction A → B in the following three scenarios.

$$k_f \gg k_r$$

$$\underbrace{K_f \times [A]}_{\text{rate of forward reaction}} = \underbrace{K_r \times [B]}_{\text{rate of reverse reaction}}$$

If the rate constants are greatly different (by many orders of magnitude), then this requires that the equilibrium concentrations of products exceed those of the reactants by the same ratio. Thus the equilibrium composition will lie strongly on the "right"; the reaction can be said to be "complete" or "quantitative".

$$k_f \ll k_r$$

$$\underbrace{K_f \times [A]}_{\text{rate of forward reaction}} = \underbrace{K_r \times [B]}_{\text{rate of reverse reaction}}$$

The rates can only be identical if the concentrations of the products are very small. We describe the resulting equilibrium as strongly favoring the left; very little product is formed. In the most extreme cases, we might even say that "the reaction does not take place".

$$k_f \approx k_r$$

If k_f and k_r have comparable values then signficant concentrations of products and reactants are present at equilibrium; we say the the reaction is "incomplete" and "reversible". The images shown in figure 8.12 offer yet another way of looking at these three cases. The plots show how the relative concentrations of the reactant and product change during the course of the reaction. The plots differ in the assumptions we make about the ratio of k_f to k_r. The equilibrium composition of the system is illustrated by the proportions of A and B in the horizontal parts of each plot where the composition remains unchanged. In each case, the two rate constants are sufficiently close in magnitude that each reaction can be considered "incomplete".

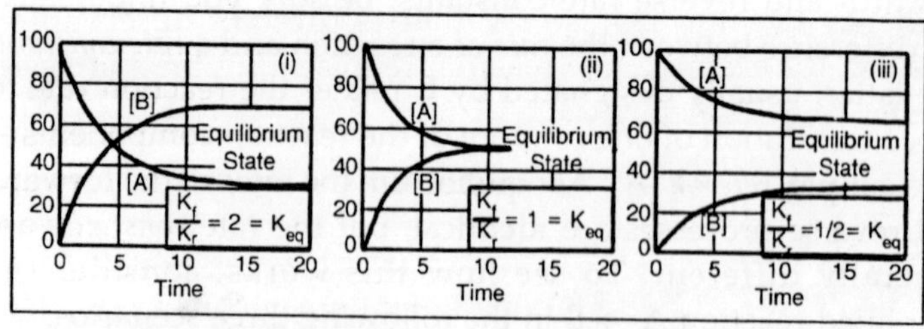

Fig. 8.12 Here For a Very Nice Online Simulation That Allows You to Explore The Effects of Your own Set of Starting Concentrations and Rate Constants on This Kind of Plot.

1. In plot *(i)* the forward rate constant is twice as large as the reverse rate constant, so product (B) is favoured, but there is sufficient reverse reaction to maintain a significant quantity of A.
2. In *(ii)*, the forward and reverse rate constants have identical magnitudes. Not surprisingly, so are the equilibrium values of [A] and [B].
3. In *(iii)*, the reverse rate constant exceeds the forward rate constant, so the eqilibrium composition is definitely "on the left".

The Law of Mass Action is thus essentially the statement that the equilibrium composition of a reaction mixture can vary according to the quantities of components that are present. This of course is just what Berthollet observed in his Egyptian salt ponds, but we now understand it to be a consequence of the dynamic nature of chemical equilibrium.

How do we know when a Reaction is at Equilibrium?

Clearly, if we observe some change taking place— a change in colour, the release of gas bubbles, the appearance of a precipitate, or the release of heat, we know the reaction is not yet at equilibrium. But the absence of any apparent change does not by itself establish that the reaction is at equilibrium. The equilibrium state is one in which not only no change in composition take place, but also one in which no energetic tendency for further change is present. Unfortunately, "tendency" is not a property that is directly observable! Consider, for example, the reaction representing the synthesis of water from its elements:

$$2\,H_2(g) + O_2(g) \rightarrow 2\,H_2O(g)$$

You can store the two gaseous reactants in the same container indefinitely without any observable change occurring. But if you create an electrical spark in the container or introduce a flame, *bang*! After you pick yourself up off the floor and remove the shrapnel from what's left of your body, you will know very well that the system was not initially at equilibrium! It happens that this particular reaction has a tremendous tendency to take place, but for reasons that we will discuss in a later chapter, nothing can happen until we "set it off" in some way— in this case by exposing the mixture to a flame or spark, or (in a more gentle way) by introducing a platinum wire, which acts as a *catalyst*. A reaction of this kind is said to be highly favoured thermodynamically, but inhibited kinetically. The similar reaction of hydrogen and iodine

$$H_2(g) + I_2(g) \rightarrow 2\,HI(g)$$

by contrast is only moderately favoured thermodynamically (and is thus incomplete), but its kinetics are both unspectacular and reasonably facile.

Some Simple Tests for the Equilibrium State

- As we explained above in the context of the law of mass action, addition or removal of one component of the reaction will affect the amounts of all the others. For example, if we add more of a reactant, we would expect to see the concentration of a product change. If this does not happen, then it is likely that the reaction is kinetically inhibited and that the system is unable to attain equilibrium.
- It is almost always the case, however, that once a reaction actually starts, it will continue on its own until it reaches equilibrium, so if we can observe the change as it occurs and see it slow down and stop, we can be reasonably certain that the system is in equilibrium. This is by far the chemist's most common criterion.
- There is one other experimental test for equilibrium in a chemical reaction, although it is really only applicable to the kind of reactions we described above as being reversible. As we shall see later, the equilibrium state of a system is always sensitive to the temperature, and often to the pressure, so any changes in these variables, however, small, will temporarily disrupt the equilibrium, resulting in an observable change in the composition of the system as it moves towards its new equilibrium state.

SUMMARY

It is especially imortant that you know the precise meanings of all the highlighted terms in the context of this topic:

- Any reaction that can be represented by a balanced chemical equation can take place, at least in principle. However, there are two important qualifications:
  1. The tendency for the change to occur may be so small that the quantity of products formed may be very low, and perhaps negligible. A reaction of this kind is said to be *thermodynamically inhibited*. The tendency for chemical change is governed solely by the properties of the reactants and products, and can be predicted by applying the laws of thermodynamics.
  2. The rate at which the reaction proceeds may be very small, or even zero, in which case we say the reaction is *kinetically inhibited*. Reaction rates depend on the *mechanism* of the reaction— that is, on what actually happens to the atoms as reactants are transformed into products. Reaction mechanisms cannot generally be predicted, and must be worked out experimentally. Also, the same reaction may have different mechansims under different conditions.
- As a chemical change proceeds, the quantities of the components on one side of the reaction equation will decrease, and those on the other side will increase. Eventually the reaction slows down and the composition of the system stops changing. At this point the reaction is in its *equilibrium state*, and no further change in composition will occur as long as the system is left undisturbed.
- For many reactions, the equilibrium state is one in which components on both sides of the equation (that is, both reactants and products) are present in significant amounts. Such a reaction is said to be *incomplete* or *reversible*.

- The equilibrium composition is independent of the direction from which it is approached; the labeling of substances as "reactants" or "products" is entirely a matter of convenience.
- The *law of mass action* states that any chemical change is a competition between a *forward* reaction (left-to-right in the chemical equation) and a *reverse* reaction. The rate of each of these processes is governed by the concentrations of the substances reacting; as the reaction proceeds, these rates approach each other and at equilibrium they become identical.
- From the above, it follows that equilibrium is a *dynamic process* in which *microscopic change* (the forward and reverse reactions) continues to occur, but*macroscopic change* (changes in the quantities of substances) is absent.
- When a chemical reaction is at equilibrium, any disturbance of the system, such as a change in temperature, or addition or removal of one of the reaction components, will "shift" the composition to a new equilibrium state. This is the only unambiguous way of verifying that a reaction is at equilibrium. The fact that the composition remains static does not in itself prove that a reaction is at equilibrium, because the change may be kinetically inhibited.

EQUILIBRIUM CONSTANT

To determine the amount of each compound that will be present at equilibrium you must know the equilibrium constant. To determine the equilibrium constant you must consider the generic equation:

$$aA + bB \leftrightarrow cC + dD$$

The upper case letters are the molar concentrations of the reactants and products. The lower case letters are the coefficients that balance the equation. Use the following equation to determine the equilibrium constant (K_c).

$$K_c = \frac{[C]^c [D]^d}{[A]^a [B]^b}$$

For example, determining the equilibrium constant of the following equation can be accomplished by using the K_c equation. Using the following equation, calculate the equilibrium constant.

$$N_2(g) + 3H_2(g) \leftrightarrow 2NH_3(g)$$

A one-liter vessel contains 1.60 moles NH_3.800 moles N_2, and 1.20 moles of H_2. What is the equilibrium constant?

$$K_c = \frac{[1.60]^2}{[.800][1.20]^3} = 29.6$$

Answer: 1.85

For a general *elementary* chemical reaction,

$$aA + bB \leftrightarrow cC + dD$$

The concentrations of the reactants and products are related to each other according to,

$$K_c = \frac{[C]^c [D]^d}{[A]^a [B]^b}$$

The number K_c is called the *equilibrium constant,* and is a function of temperature only (*i.e.*, its numerical value doesn't change unless the temperature changes–we'll hold the temperature constant for now). Note the word *elementary,* more on that later; for now, all reactions are elementary. The stoichiometric coefficients a, b, c and d show up as powers of the corresponding reactants (and products–just call them all reactants from now on). For our example reaction,

$$A + B \leftrightarrow C$$

the equilibrium constant is defined,

$$K_c = \frac{[C]}{[A][B]}$$

Notice that the units of K_c in this case are M^{-1}= L/mol.

There is another useful definition of the equilibrium constant based on pressure rather than concentration. The ideal gas law reads,

$$PV = nRT$$

Here P is the total pressure. In the case of several components, each has a partial pressure, all of which sum up to the total pressure:

$$P = P_A + P_a + P_c$$

For each component, we can write the ideal gas law (putting a subscript where applicable)

$$P_A V = n_A RT \Rightarrow \frac{n_A}{V} = [A] = \frac{P_A}{RT}$$

This works for the other components, too, and gives us the relation between the concentrations and the partial pressures. If we plug in the partial pressures in the definition of K_c above, we get,

$$K_c = \frac{[C]}{[A][B]} = \frac{P_c}{RT}\frac{RT}{P_A}\frac{RT}{P_B} = \frac{P_c}{P_A P_B}(RT) = K_P(RT)$$

In the more general case where the reaction is,

$$aA + bB \leftrightarrow cC + dD$$

And K_c is,

$$K_c = \frac{[C]^c [D]^d}{[A]^a [B]^b}$$

Then the new equilibrium constant becomes,

$$K_P = K_c (RT)^{(c+d)-(a+b)} = K_c (RT)^{\Delta n}$$

Where,

$$\Delta n = (c+d)-(a+b)$$

Is the change in number of moles in the gas phase. These numbers come from the balanced equation and are basically the sum of the stoichiometric coefficients of the products minus the sum of the stoichiometric coefficients of the reactants.

Problems using K_P instead of K_c are solved the same way, except we are looking for the partial pressures P_X instead of the concentrations [X], X = A, B, C, etc. is particularly useful for gas phase problems since the pressure is usually more convenient to measure than the concentration. Also, there is a direct relationship between K_P and the Gibbs energy of the reaction, $\Delta G^0{}_{rxn}$.

LE CHATELIER'S PRINCIPLE

Le Chatelier's principle states that when a system in chemical equilibrium is disturbed by a change of temperature, pressure, or a concentration, the system shifts in equilibrium composition in a way that tends to counteract this change of variable.

The three ways that Le Chatelier's principle says you can affect the outcome of the equilibrium are as follows:

1. Changing concentrations by adding or removing products or reactants to the reaction vessel.
2. Changing partial pressure of gaseous reactants and products.
3. Changing the temperature.

These actions change each equilibrium differently, therefore you must determine what needs to happen for the reaction to get back in equilibrium.

Example involving change of concentration:

In the equation,

$$2NO_{(g)} + O_{2(g)} \leftrightarrow 2NO_{2(g)}$$

- If you add more $NO_{(g)}$ the equilibrium shifts to the right producing more $NO_{2(g)}$
- If you add more $O_{2(g)}$ the equilibrium shifts to the right producing more $NO_{2(g)}$
- If you add more $NO_{2(g)}$ the equilibrium shifts to the left producing more $NO_{(g)}$ and $O_{2(g)}$

Example involving pressure change:

In the equation,

$$2SO_{2(g)} + O_{2(g)} \leftrightarrow 2SO_{3(g)},$$

an increase in pressure will cause the reaction to shift in the direction that reduces pressure, that is the side with the fewer number of gas molecules. Therefore an increase in pressure will cause a shift to the right, producing more product. (A decrease in volume is one way of increasing pressure.)

Example involving temperature change:

In the equation,

$$N_{2(g)} + 3H_{2(g)} \leftrightarrow 2NH_3 + 91.8\ kJ,$$

an increase in temperature will cause a shift to the left because the reverse reaction uses the excess heat. An increase in forward reaction would produce even more heat since the forward reaction is exothermic. Therefore the shift caused by a change in temperature depends upon whether the reaction is exothermic or endothermic. If a reaction is at equilibrium and we alter the conditions so as to create a new*equilibrium state,* then the composition of the system will tend to change until that new equilibrium state is attained. (We say "tend to change" because if the reaction is *kinetically inhibited,* the change may be too slow to observe or it may never take place.)

In 1884, the French chemical engineer and teacherHenri Le Châtelier (1850–1936) showed that in every such case, the new equilibrium state is one that partially reduces the effect of the change that brought it about.This law is known to every Chemistry student as the *Le Châtelier principle.* His original formulation was somewhat complicated, but a reasonably useful paraphrase of it reads as follows: To see how this works (and you *must* do so, as this is of such fundamental importance that you simply cannot do any meaningful chemistry without a thorough working understanding of this principle), look again at the hydrogen iodide dissociation reaction

$$2HI \rightarrow H_2 + I_2$$

Consider an arbitrary mixture of these three components at equilibrium, and assume that we inject more hydrogen gas into the container. Because the H_2 concentration now exceeds its new equilibrium value, the system is no longer in its equilibrium state, so a net reaction now ensues as the system moves to the new state.

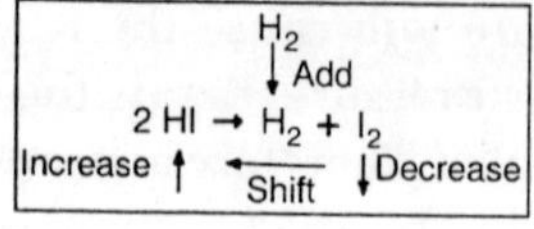

Fig. 8.13

The Le Châtelier principle states that the net reaction will be in a direction that tends to reduce the effect of the added H_2. This can occur if some of the H_2 is consumed by reacting with I_2 to form more HI; in other words, a net reaction occurs in the reverse direction. Chemists usually simply say that "the equilibrium shifts to the left".

To get a better idea of how this works, carefully examine the diagram below which follows the concentrations of the three components of this reaction as they might change in time (the time scale here will typically be about an hour):

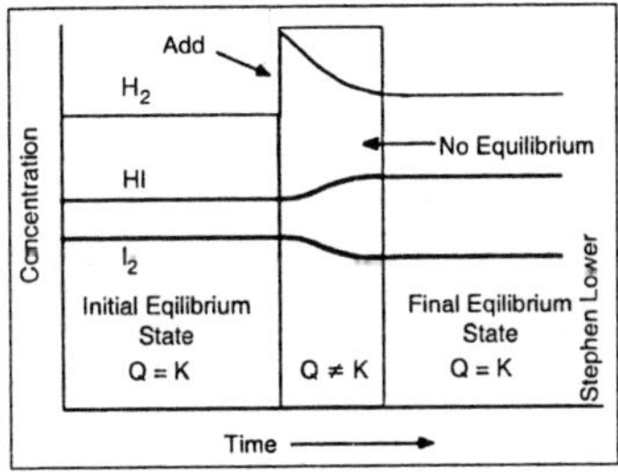

Fig. 8.14

DISRUPTION AND RESTORATION OF EQUILIBRIUM

At the left, the concentrations of the three components do not change with time because the system is at equilibrium. We then add more hydrogen to the system, disrupting the equilibrium. A net reaction then ensues that moves the system to a new equilibrium state (right) in which the quantity of hydrogen iodide has increased; in the process, some of the I_2 and H_2 are consumed.

Notice that the new equilibrium state contains more hydrogen than did the initial state, but not as much as was added; as the le Châtelier principle predicts, the change we made (addition of H_2) has been partially counteracted by the "shift to the right".

The following tabLe Contains several examples showing how changing the quantity of a reaction component can shift an established equilibrium.

System	Change	Result
$CO_2 + H_2 \rightarrow H_2O(g) + CO$	a drying agent is added to absorb H_2O	**Shift to the right.** Continuous removal of a product will force any reaction to the right
$H_2(g) + I_2(g) \rightarrow 2HI(g)$	Some nitrogen gas is added	**No change;** N_2 is not a component of this reaction system.
$NaCl(s) + H_2SO_4(l) \rightarrow Na_2SO_4(s) + HCl(g)$	reaction is carried out in an open container	Because HCl is a gas that can escape from the system, the reaction is forced to the **right**. This is the basis for the commercial production of hydrochloric acid.
$H_2O(l) \rightarrow H_2O(g)$	water evaporates from an open container	Continuous removal of water vapor forces the reaction to the **right**, so equilibrium is never achieved.
$HCN(aq) \rightarrow H^+(aq) + CN^-(aq)$	the solution is diluted	Shift to **right**; the product $[H^+][CN^-]$ diminishes more rapidly than does [HCN].
$AgCl(s) \rightarrow Ag^+(aq) + Cl^-(aq)$	some NaCl is added to the solution	Shift to **left** due to increase in Cl^- concentration. This is known as the common ion effect on solubility.
$N_2 + 3\ H_2 \rightarrow 2\ NH_3$	a catalyst is added to speed up this reaction	**No change**. Catalysts affect only the rate of a reaction; the have no effect at all on the composition of the equilibrium state.

HOW DO CHANGES IN TEMPERATURE AFFECT EQUILIBRIA

Virtually all chemical reactions are accompanied by the liberation or uptake of heat. If we regard heat as a "reactant" or "product" in an endothermic or exothermic reaction respectively, we can use the Le Chatelier principle to predict the direction in which an increase or decrease in temperature will shift the equilibrium state. Thus for the oxidation of nitrogen, an endothermic process, we can write,

$$[\text{heat}] + N_2 + O_2 \rightarrow 2\ NO$$

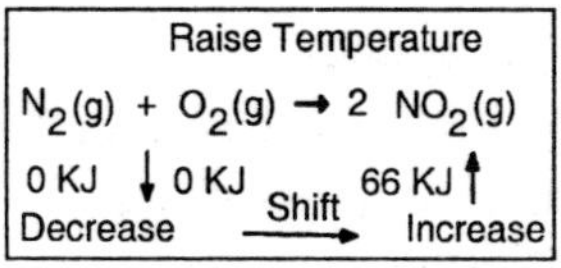

Raise Temperature		
$N_2(g)$ + $O_2(g)$	→	2 $NO_2(g)$
0 KJ ↓ 0 KJ	Shift	66 KJ ↑
Decrease	⟶	Increase

Suppose this reaction is at equilibrium at some temperature T_1 and we raise the temperature to T_2. The Le Châtelier principle tells us that a net reaction will occur in the direction that will partially counteract this change. Since the reaction is endothermic, a shift of the equilibrium to the right will take place.

Nitric oxide, the product of this reaction, is a major air pollutant which initiates a sequence of steps leading to the formation of atmospheric smog. Its formation is an unwanted side reaction which occurs when the air (which is introduced into the combustion chamber of an engine to supply oxygen) gets heated to a high temperature. Designers of internal combustion engines now try, by various means, to limit the temperature in the combustion region, or to restrict its highest-temperature part to a small volume within the combustion chamber.

HOW DO CHANGES IN PRESSURE AFFECT EQUILIBRIA

You will recall that if the pressure of a gas is reduced, its volume will increase; pressure and volume are inversely proportional. With this in mind, suppose that the reaction

$$2\ NO_2(g) \rightarrow N_2O_4(g)$$

is in equilibrium at some arbitrary temperature and pressure, and that we double the pressure, perhaps by compressing the mixture to a smaller volume. From the Le Châtelier principle we know that the equilibrium state will change to one that tends to counteract the increase in pressure. This can occur if some of the NO_2 reacts to form more of the dinitrogen tetroxide, since two moles of gas are being removed from the system for every mole of N_2O_4 formed, thereby decreasing the total volume of the system. Thus increasing the pressure will shift this equilibrium to the right.

It is important to understand that changing the pressure will have a significant effect only on reactions in which there is a change in the number of moles of gas.

For the above reaction, this change

$$\Delta n_g = (n_{products} - n_{reactants}) = 1-2 = -1.$$

In the case of the nitrogen oxidation reaction,

$N_2 + O_2 \rightarrow 2\,NO$, $\Delta n_g = 0$ and pressure will have no effect.

The volumes of solids and liquids are hardly affected by the pressure at all, so for reactions that do not involve gaseous substances, the effects of pressure changes are ordinarily negligible. Exceptions arise under conditions of very high pressure such as exist in the interior of the Earth or near the bottom of the ocean. A good example is the dissolution of calcium carbonate,

$$CaCO_3(s) \rightarrow Ca^{2+} + CO_3^{2-}.$$

There is a slight decrease in the volume when this reaction takes place, so an increase in the pressure will shift the equilibrium to the right, with the results that calcium carbonate becomes more soluble at higher pressures.

The skeletons of several varieties of microscopic organisms that inhabit the top of the ocean are made of $CaCO_3$, so there is a continual rain of this substance towards the bottom of the ocean as these organisms die. As a consequence, the floor of the Atlantic ocean is covered with a blanket of calcium carbonate.

This is not true for the Pacific ocean, which is deeper; once the skeletons fall below a certain depth, the higher pressure causes them to dissolve. Some of the seamounts (undersea mountains) in the Pacific extend above the solubility boundary so that their upper parts are covered with $CaCO_3$ sediments.

The effect of pressure on a reaction involving substances whose boiling points fall within the range of commonly encountered temperature will be sensitive to the states of these substances at the temperature of interest. For reactions involving gases, only changes in the partial pressures of those gases directly involved in the reaction are important; the presence of other gases has no effect.

Problem Example

The commercial production of hydrogen is carried out by treating natural gas with steam at high temperatures and in the presence of a catalyst ("steam reforming of methane"):

$$CH_4 + H_2O \rightarrow CH_3OH + H_2$$

Given the following boiling points: CH_4 (methane) = –161°C, H_2O = 100°C, CH_3OH = 65°, H_2 = –253°C, predict the effects of an increase in the total pressure on this equilibrium at 50°, 75° and 120°C.

Solution: Calculate the change in the moles of gas for each process:

Temp	Equation	Δn_g	Shift
50°	$CH_4(g) + H_2O(l) \rightarrow CH_3OH(l) + H_2(g)$	0	none
75°	$CH_4(g) + H_2O(l) \rightarrow CH_3OH(g) + H_2(g)$	+1	to left
120°	$CH_4(g) + H_2O(g) \rightarrow CH_3OH(g) + H_2(g)$	0	none

WHAT IS THE HABER PROCESS AND WHY IS IT IMPORTANT

The Haber process for the synthesis of ammonia is based on the exothermic reaction:

$$N_2(g) + 3\,H_2(g) \rightarrow 2\,NH_3(g) \quad \Delta H = -92 \text{ kJ/mol}$$

The Le Chatelier principle tells us that in order to maximize the amount of product in the reaction mixture, it should be carried out at high pressure and low temperature. However, the lower the temperature, the slower the reaction (this is true of virtually all chemical reactions.) As long as the choice had to be made between a low yield of ammonia quickly or a high yield over a long period of time, this reaction was infeasible economically.

Nitrogen is available for free, being the major component of air, but the strong triple bond in N_2 makes it extremely difficult to incorporate this element into species such as NO_3^- and NH_4^+ which serve as the starting points for the wide variety of nitrogen-containing compounds that are essential for modern industry. This conversion is known as *nitrogen fixation,* and because nitrogen is an essential plant nutrient, modern intensive agriculture is utterly dependent on huge amounts of fixed nitrogen in the form of fertilizer. Until around 1900, the major source of fixed nitrogen was the $NaNO_3$ found in extensive deposits in South America. Several chemical processes for obtaining nitrogen compounds were developed in the early 1900's, but they proved too inefficient to meet the increasing demand.

$CH_4 + H_2O \rightarrow CO + 3H_2$ formation of synthesis gas from methane

$CO + H_2O \rightarrow CO_2 + H_2$ *shift reaction* carried out in reformer

The Haber-Bosch process is considered the most important chemical synthesis developed in the 20th century. Besides its scientific importance as the first large-scale application of the laws of chemical equilibrium, it has had tremendous economic and social impact; without an inexpensive source of fixed nitrogen, the intensive crop production required to feed the world's growing population would have been impossible. Haber was awarded the 1918 Nobel Prize in Chemistry in recognition of his

THE LE CHATELIER PRINCIPLE IN PHYSIOLOGY

Many of the chemical reactions that occur in living organisms are regulated through the Le Châtelier principle.

OXYGEN TRANSPORT BY THE BLOOD

Few of these are more important to warm-blooded organisms than those that relate to aerobic respiration, in which oxygen is transported to the cells where it is combined with glucose and metabolized to carbon dioxide, which then moves back to the lungs from which it is expelled.

$$\text{Hemoglobin} + O_2 \rightleftharpoons \text{Oxyhemoglobin}$$

The partial pressure of O_2 in the air is 0.2 atm, sufficient to allow these molecules to be taken up by hemoglobin (the red pigment of blood) in which it becomes loosely bound in a complex known as oxyhemoglobin. At the ends of the capillaries which deliver the blood to the tissues, the O_2 concentration is reduced by about 50% owing to its consumption by the cells. This shifts the equilibrium to the left, releasing the oxygen so it can diffuse into the cells.

Much more detail about the mechanism by which hemoglobin transfers oxygen from the lungs to tissues, and then carries carbon dioxide and hydrogen ions back to the lungs, can be found on this U. of Virginia page.

MAINTENCE OF BLOOD PH

Carbon dioxide reacts with water to form a weak acid H_2CO_3 which would cause the blood pH to fall to dangerous levels if it were not promptly removed as it is excreted by the cells. This is accomplished by combining it with carbonate ion through the reaction,

$$H_2CO_3 + CO_3^{2-} \rightleftharpoons 2HCO_3^-$$

which is forced to the right by the high local CO_2 concentration within the tissues. Once the hydrogen carbonate (bicarbonate) ions reach the lung tissues where the CO_2 partial pressure is much smaller, the reaction reverses and the CO_2 is expelled.

CARBON MONOXIDE POISONING

Carbon monoxide, a product of incomplete combustion that is present in automotive exhaust and cigarette smoke, binds to hemoglobin 200 times more tightly than does O_2. This blocks the uptake and transport of oxygen by setting up a competing equilibrium

$$O_2\text{-hemoglobin} \rightleftharpoons \text{hemoglobin} \rightleftharpoons \text{CO-hemoglobin}$$

Air that contains as little as 0.1 per cent carbon monoxide can tie up about half of the hemoglobin binding sites, reducing the amount of O_2 reaching the tissues to fatal levels. Carbon monoxide poisoning is treated by administration of pure O_2 which promotes the shift of the above equilibrium to the left. This can be made even more effective by placing the victim in a hyperbaric chamber in which the pressure of O_2 can be made greater than 1 atm.

Make sure you thoroughly understand the following essential ideas which have been presented above. It is especially imortant that you know the precise meanings of all the highlighted terms in the context of this topic.

- A system in its *equilibrium state* will remain in that state indefinitely as long as it is undisturbed. If the equilibrium is destroyed by subjecting the system to a change of pressure, temperature, or the number of moles of a substance, then a net reaction will tend to take place that moves the system to a new equilibrium state. *Le Châtelier's principle* says that this net reaction will occur in a direction that partially offsets the change.
- The Le Châtelier Principle has practical effect only for reactions in which signficant quantities of both reactants and products are present at equilibrium—that is, for reactions that are *thermodynamically reversible.*
- Addition of more product substances to an equilibrium mixture will shift the equilibrium to the left; addition of more reactant substances will shift it to the right. These effects are easily explained in terms of competing forward- and reverse reactions—that is, by the *law of mass action.*
- If a reaction is *exothermic* (releases heat), an increase in the temperature will force the equilibrium to the left, causing the system to absorb heat and thus partially ofsetting the rise in temperature. The opposite effect occurs for*endothermic* reactions, which are shifted to the right by rising temperature.
- The effect of pressure on an equilibrium is significant only for reactions which involve different numbers of

moles of gases on the two sides of the equation. If the number of moles of gases increases, than an increase in the total pressure will tend to initiate a reverse reaction that consumes some the products, partially reducing the effect of the pressure increase.

- The classic example of the practical use of the Le Châtelier principle is the *Haber-Bosch process* for the synthesis of ammonia, in which a balance between low temperature and high pressure must be found.

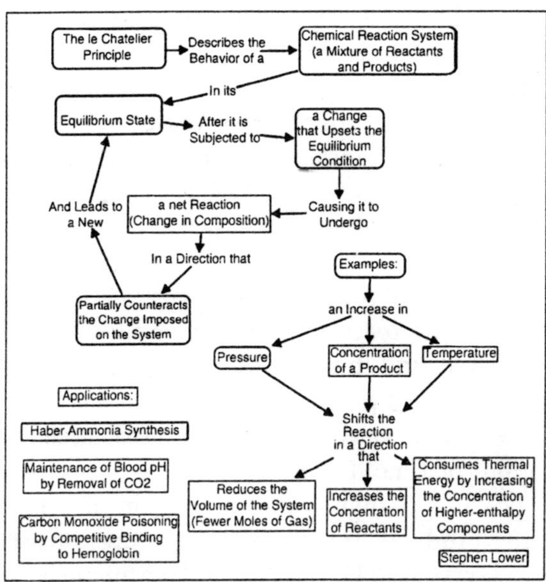

Fig. 8.15

Q AND *K*: WHAT'S THE DIFFERENCE

Given a chemical reaction system and its equilibrium constant, together with the concentrations or pressures of the reactants and products actually present, we often need to predict whether, and in which direction, a net change in composition will tend to take place. The concept of the reaction quotient makes this very easy to do, but the similarity of the algebraic forms of K and Q can easily lead to confusion.

WHAT IS THE EQUILIBRIUM QUOTIENT

We defined the *equilibrium expression* for the reaction

$$a\,\mathrm{A} + b\,\mathrm{B} \rightarrow c\,\mathrm{C} + d\,\mathrm{D} \text{ as}$$

$$\frac{[\mathrm{C}]^c\,[\mathrm{D}]^d}{[\mathrm{A}]^a\,[\mathrm{B}]^b}$$

In the general case in which the concentrations can have any arbitrary values (including zero), this expression is called the *reaction quotient*(the term *equilibrium quotient* is also commonly used.) and its value is denoted by Q (or Q_c or Q_p if we wish to emphasize that the terms represent molar concentrations or partial pressures.) If the terms correspond to *equilibrium* concentrations, then the above expression is called the *equilibrium constant* and its value is denoted by K (or K_c orK_p.)

K is thus the special value that Q has when the reaction is at equilibrium

The value of Q in relation to K serves as an index how the composition of the reaction system compares to that of the equilibrium state, and thus it indicates the direction in which any net reaction must proceed.

For example, if we combine the two reactants A and B at concentrations of 1 mol L^{-1} each, the value of Q will be 0÷1=0. The only possible change is the conversion of some of these reacants into products. If instead our mixture consists only of the two products C and D, Q will be indeterminately large (1÷0) and the only possible change will be in the reverse direction.

It is easy to see (by simple application of the le Châtelier principle) that the ratio of Q/K immediately tells us whether, and in which direction, a net reaction will occur as the system moves towards its equilibrium state. A schematic view of this relationship is shown below:

$Q = \frac{[\text{Products}]}{[\text{Reactants}]}$	$K = Q = \frac{[\text{Products}]}{[\text{Reactants}]}$	$Q = \frac{[\text{Products}]}{[\text{Reactants}]}$
$Q > K$	$Q = K$	$Q < K$
←	↔	→
Net Reaction to Left	No net Reaction	Net Reaction to Right

More formally, we simply look at which of the following describes the relative values of *Q* and *K*:

Q/K	
> 1	Product concentration too high for equilibrium; net reaction proceeds to left.
= 1	System is at equilibrium; no net change will occur.
< 1	Product concentration too low for equilibrium; net reaction proceeds to right.

It is very important that you be able to work out these relations for yourself, not by memorizing them, but from the definitions of *Q* and *K*.

Problem Example

The equilibrium constant for the oxidation of sulfur dioxide is $K_p = 0.14$ at 900 K.

$$2\ SO_2(g) + O_2(g) \rightarrow 2\ SO_3(g)$$

If a reaction vessel is filled with SO_3 at a partial pressure of 0.10 atm and with O_2 and SO_2 each at a partial pressure of 0.20 atm, what can you conclude about whether, and in which direction, any net change in composition will take place?

Solution:

The value of the equilibrium quotient *Q* for the initial conditions is,

$$Q = \frac{(PSO_3)^2}{(PO_2)(PSO_2)^2} = \frac{(0.10\,atm)^2}{(0.20\,atm)(0.20\,atm)^2} = 1.25\,atm^{-1}$$

Since *Q*>*K*, the reaction is not at equilibrium, so a net change will occur in a direction that decreases *Q*. This can only occur if some of the SO_3 is converted back into products. In other words, the reaction will "shift to the left".

A VISUAL WAY OF THINKING ABOUT *Q* AND *K*

The formal definitions of *Q* and *K* are quite simple, but they are of limited usefulness unless you are able to relate them to real chemical situations.

The following diagrams illustrate the relation between Q and K from various standpoints. Take some time to study each one carefully, making sure that you are able to relate the description to the illustration.

Example: Dissociation of Dinitrogen Tetroxide

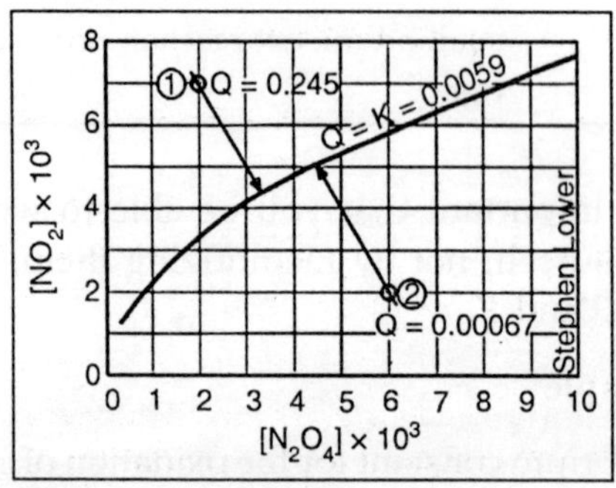

Fig. 8.16

For the reaction $N_2O_4 \rightarrow 2\ NO_2$, $K_c = 0.0059$ at 298K.

This equilibrium condition is represented by the red curve that passes through all points on the graph that satisfy the requirement that,

$$Q = [NO_2]^2 / [N_2O_4] = 0.0059.$$

There are of course an infinite number of possible Q's of this system within the concentration boundaries shown on the plot. Only those points that fall on the red line correspond to *equilibrium states* of this system (those for which $Q = K$). The line itself is a plot of $[NO_2]$ that we obtain by rearranging the equilibrium expression $[NO_2] = ([N_2O_4]K)^{0.5}$.

If the system is initially in a non-equilibrium state, its composition will tend to change in a direction that moves it to one that is on the line.

Two such non-equilibrium states are shown. The state indicated by has 1. $Q > K$, so we would expect a net reaction that reduces Q by converting some of the NO_2 into N_2O_4; in other words, the equilibrium "shifts to the left". Similarly, in state 2. $Q < K$, indicating that the forward reaction will occur.

The arrows in the figure 8.16 indicate the successive values that Q assumes as the reaction moves closer to equilibrium. The slope of the line reflects the stoichiometry of the equation. In this case, one mole of reactant yields two moles of products, so the slopes have an absolute value of 2:1.

Example 2: Phase-Change Equilibrium

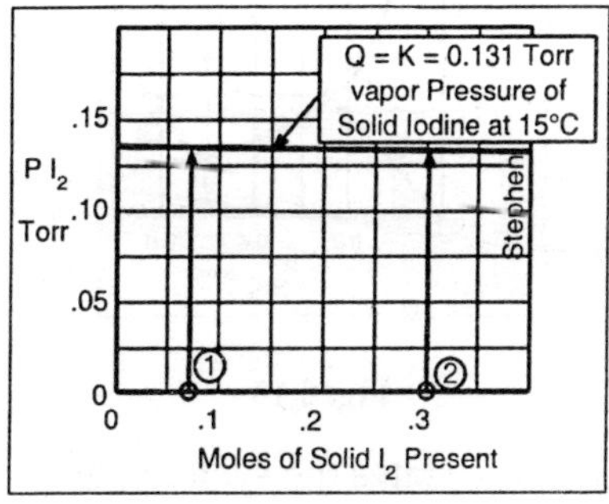

Fig. 8.17

One of the simplest equilibria we can write is that between a solid and its vapour. In this case (as will be explained more fully in the next lesson), the equilibrium constant is just the vapour pressure of the solid. Thus for the process,

$$I_2(s) \rightarrow I_2(g)$$

all possible equilibrium states of the system lie on the horizontal red line and is independent of the quantity of solid present (as long as there is at least enough to supply the relative tiny quantity of vapour.)

So adding various amounts of the solid to an empty closed vessel (states 1. and 2.) causes a gradual buidup of iodine vapour. Because the equilibrium pressure of the vapour is so small, the amount of solid consumed in the process is negligible, so the arrows go straight up and all lead to the same equilibrium vapour pressure.

Example: Heterogeneous Chemical Reaction

The decomposition of ammonium chloride is a common example of a heterogeneous (two–phase) equilibrium. Solid ammonium chloride has a substantial vapour pressure even at room temperature:

$$NH_4Cl(s) \rightarrow NH_3(g) + HCl(g)$$

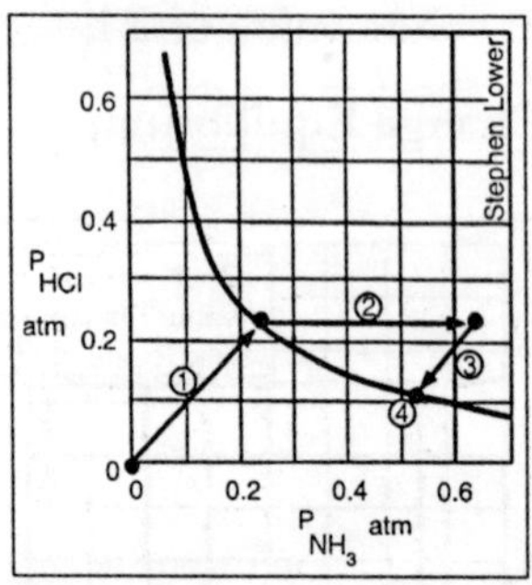

Fig. 8.18

Arrow 1. traces the states the system passes through when solid NH_4Cl is placed in a closed container. Arrow 2. represents the addition of ammonia to the equilibrium mixture; the system responds by following the path 3. back to a new equilibrium state 4. which, as the Le Châtelier principle predicts, contains a smaller quantity of ammonia than was added. The unit slopes of the paths 1. and 2. reflect the 1:1 stoichiometry of the gaseous products of the rection.

It is especially imortant that you know the precise meanings of all the highlighted terms in the context of this topic:

- When arbitrary quantities of the different *components* of a *chemical reaction system* are combined, the overall system composition will not likely correspond to the equilibrium composition. As a result, a net change in composition ("a shift to the right or left") will tend to take place until the *equilibrium state* is attained.
- The status of the reaction system in regard to its equilibrium state is characterised by the value of the *equilibrium expression* whose formulation is defined by

the coefficients in the balanced reaction equation; it may be expressed in terms of concentrations, or in the case of gaseous components, as partial pressures.

- The various terms in the equilibrium expression can have any arbitrary value (including zero); the value of the equilibrium expression itself is called the*reaction quotient Q*.
- If the concentration or pressure terms in the equilibrium expression correspond to the equilibrium state of the system, then Q has the special value K, which we call the *equilibrium constant*.
- The ratio of Q/K (whether it is 1, >1 or <1) thus serves as an index of how far the system is from its equilibrium composition, and its value indicates the direction in which the net reaction must proceed in order to reach its equilibrium state.
- When $Q = K$, then the equilibrium state has been reached, and no further net change in composition will take place as long as the system remains undisturbed.

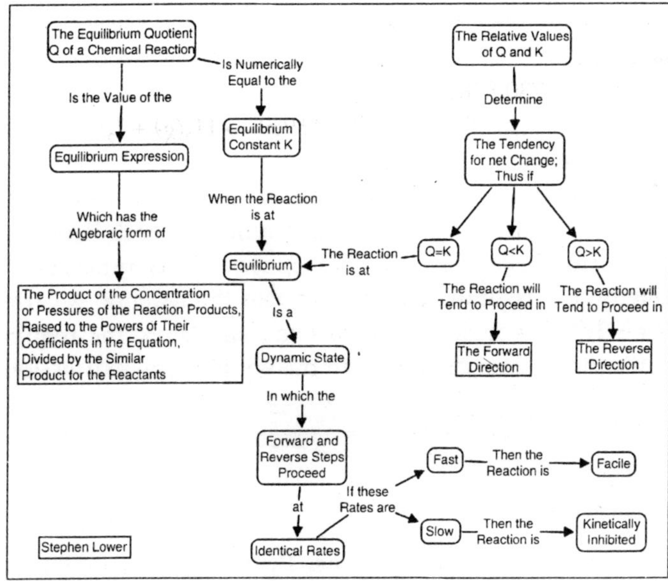

Fig. 8.19

EQUILIBRIUM CALCULATIONS

MEASURING AND CALCULATING EQUILIBRIUM CONSTANTS

Clearly, if the concentrations or pressures of all the components of a reaction are known, then the value of *K* can be found by simple substitution. Observing individual concentrations or partial pressures directly may be not always be practical, however. If one of the components is coloured, the extent to which it absorbs light of an appropriate wavelength may serve as an index of its concentration. Pressure measurements are ordinarily able to measure only the total pressure of a gaseous mixture, so if two or more gaseous products are present in the equilibrium mixture, the partial pressure of one may need to be inferred from that of the other, taking into account the stoichiometry of the reaction.

Problem Example

In an experiment carried out by Taylor and Krist (*J. Am. Chem. Soc.* 1941: 1377), hydrogen iodide was found to be 22.3% dissociated at 730.8°K.

Calculate K_c for $2\ HI(g) \rightarrow H_2(g) + I_2$.

Solution:

No explicit molar concentrations are given, but we do know that for every *n*moles of HI, 0.223*n* moles of each product is formed and (1–0.233)*n* = 0.777*n* moles of HI remains. For simplicity, we assume that *n*=1 and that the reaction is carried out in a 1.00-L vessel, so that we can substitute the required concentration terms directly into the equilibrium expression.

$$K_c = \frac{[H_2][I_2]}{[HI]^2} = \frac{(0.233)^2}{(0.777)^2} = 0.12$$

Problem Example

Ordinary white phosphorus, P_4, forms a vapour which dissociates into diatomic molecules at high temperatures:

$$P_4(g) \rightarrow 2\ P_2(g)$$

A sample of white phosphorus, when heated to 1000°C, formed a vapour having a total pressure of 0.20 atm and a density of 0.152 g L^{-1}. Use this information to evaluate the equilibrium constant K_p for this reaction.

Solution: Before worrying about what the density of the gas mixture has to do with K_p, start out in the usual way by laying out the information required to express K_p in terms of an unknown x.

	P_4	$2\ P_2$	
Initial moles:	1	$1-x$	Since K is independent of the number of moles, assume the simplest case.
Moles at equilibrium:	$1-x$	$2x$	x is the fraction of P_4 that dissociates.
Eq. mole fractions:	$\left(\frac{1-x}{1+x}\right)$	$\left(\frac{2x}{1+x}\right)$	The denominator is the total number of moles. $(1-x)+2x = 1+x$.
Eq. partial pressures:	$\left(\frac{1-x}{1+x}\right)\times 0.2$	$\left(\frac{2x}{1+x}\right)\times 0.2$	Partial pressure is the mole fraction times the total pressure.

Expressing the equilibrium constant in terms of x gives,

$$K_P = \frac{p_{P2}^2}{P_{P4}} = \frac{\left(\frac{1-x}{1+x}\right)^2}{\left(\frac{2x}{1+x}\right)} = \frac{(1-x)^2}{2x}$$

Now we need to find the dissociation fraction x of P_4, and at this point we hope you remember those gas laws that you were told you would be needing later in the course! The density of a gas is directly proportional to its molecular weight, so you need to calculate the densities of pure P_4 and pure P_2 vapors under the conditions of the experiment. One of these densities will be greater than 0.152 gL^{-1} and the other will be smaller; all you need to do is to find where the measured density falls in between the two limits, and you will have the dissociation fraction. The molecular weight of phosphorus is 31.97, giving a molar mass of 127.9 g for P_4. This mass must be divided by the volume to find the density; assuming ideal gas behaviour, the volume of 127.9 g (1 mole) of P_4 is given by RT/P, which works out to 522 L (remember to use the absolute temperature here.) The density of pure P_4 vapour under the conditions of the experiment is then

$d = m/V = (128 \text{ g mol}^{-1}) \times = (522 \text{ L mol}^{-1}) = 0.245 \text{ g L}^{-1}$.

The density of pure P_2 would be half this, or 0.122 g L^{-1}. The difference between these two limiting densities is 0.123 g L^{-1}, and the difference between the density of pure P_4 and that of the equilibrium mixture is (.245 –.152) g L^{-1}or 0.093 g L^{-1}. The ratio 0.093 0.123 = 0.76 is therefore the fraction of P_4that remains and its fractional dissociation is (1–0.76) = 0.24. Substituting into the equilibrium expression gives K_p = 1.2.

PREDICTING EQUILIBRIUM CONCENTRATIONS

This is by far the most common kind of equilibrium problem you will encounter: starting with an arbitrary number of moles of each component, how many moles of each will be present when the system comes to equilibrium?

The principal source of confusion and error for beginners relates to the need to determine the values of several unknowns (a concentration or pressure for each component) from a single equation, the equilibrium expression.

The key to this is to make use of the stoichiometric relationships between the various components, which usually allow us to express the equilibrium composition in terms of a single variable.

The easiest and most error-free way of doing this is adopt a systematic approach in which you create and fill in a small table as shown in the following problem example. You then substitute the equilibrium values into the equilibrium constant expression, and solve it for the unknown.

This very often involves solving a quadratic or higher-order equation. Quadratics can of course be solved by using the familiar quadratic formula, but it is often easier to use an algebraic or graphical approximation, and for higher-order equations this is the only practical approach. There is almost never any need to get an exact answer, since the equilibrium constants you start with are rarely known all that precisely anyway.

Problem Example

Phosgene ($COCl_2$) is a poisonous gas that dissociates at high temperature into two other poisonous gases, carbon monoxide and chlorine. The equilibrium constant $K_p = 0.0041$ at 600°K. Find the equilibrium composition of the system after 0.124 atm of $COCl_2$ is allowed to reach equilibrium at this temperature.

Solution: Start by drawing up a table showing the relationships between the components:

	$COCl_2$	CO	Cl_2
Initial pressures:	0.124 atm	0	0
Change:	–x	+x	+x
Equilibrium pressures:	0.124–x	x	x

Substitution of the equilibrium pressures into the equilibrium expression gives,

$$\frac{x^2}{0.124 - x} = 0.0041$$

This expression can be rearranged into standard polynomial form $x^2 + .0041\,x - 0.00054 = 0$ and solved by the quadratic formula, but we will simply obtain an approximate solution by iteration. Because the equilibrium constant is small, we know that x will be rather small compared to 0.124, so the above relation can be approximated by,

$$\frac{x^2}{0.124} = 0.0041$$

which gives $x = 0.0225$. To see how good this is, substitute this value of x into the denominator of the original equation and solve again:

$$\frac{x^2}{0.124 - 0.0225} = \frac{x^2}{0.102} = 0.0041$$

This time, solving for x gives 0.0204. Iterating once more, we get,

$$\frac{x^2}{0.124-0.0204}=\frac{x^2}{0.104}=0.0041$$

and x = 0.0206 which is sufficiently close to the previous to be considered the final result. The final partial pressures are then 0.104 atm for $COCl_2$, and 0.0206 atm each for CO and Cl_2.

Problem Example

The gas-phase dissociation of phosphorus pentachloride to the trichloride has Kp = 3.60 at 540°C:

$$PCl_5 \rightarrow PCl_3 + Cl_2$$

What will be the partial pressures of all three components if 0.200 mole of PCl_5 and 3.00 moles of PCl_3 are combined and brought to equilibrium at this temperature and at a total pressure of 1.00 atm?

Solution: As always, set up a table showing what you know (first two rows) and then expressing the equilibrium quantities:

	PCl_5	PCl_3	Cl_2
Initial moles:	0.200	3.00	0
Change:	–x	+x	+x
Equilibrium moles:	0.200 – x	3.00 + x	x
Eq. partial pressures:	$\left(\frac{2.00-x}{3.20+x}\right)$	$\left(\frac{3.00+x}{3.20+x}\right)$	$\left(\frac{x}{3.20+x}\right)$

The partial pressures in the bottom row were found by multiplying the mole fraction of each gas by the total pressure: $P_i = X_i P_t$. The term in the denominator of each mole fraction is the total number of moles of gas present at equilibrium: $(0.200–x) + (3.00 + x) + x = 3.20 + x$. Substituting the equilibrium partial pressures into the equilibrium expression, we have,

$$\frac{(3.00+x)(x)}{(0.200-x)(3.20+x)}=3.60$$

whose polynomial form is $4.60\,x^2 + 13.80\,x - 2.304 = 0$.

Plotting this on a graphical calculator yields x = 0.159 as the positive root:

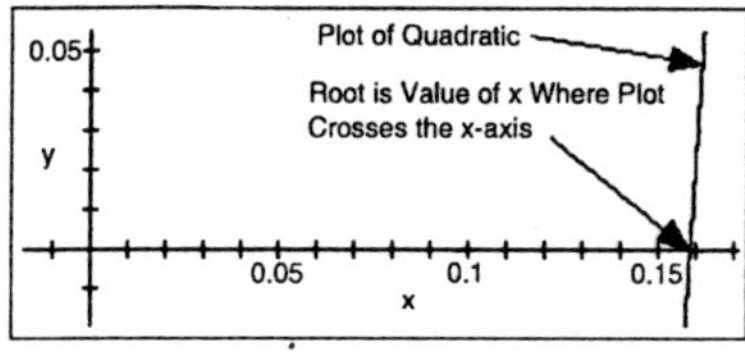

Fig. 8.20

Substitution of this root into the expressions for the equilibrium partial pressures in the table yields the following values: $P(PCl_5) = 0.012$ atm, $P(PCl_3) = 0.94$ atm, $P(Cl_2) = 0.047$ atm.

EFFECTS OF DILUTION ON EQUILIBRIUM

In this the LeChâtelier principle, it was mentioned that diluting a weak acid such as acetic acid CH_3COOH ("HAc") will shift the dissociation equilibrium to the right:

$$HAc + H_2O \rightarrow H_3O^+ + Ac^-$$

Thus a 0.10*M* solution of acetic acid is 1.3% ionized, while in a 0.01M solution, 4.3% of the HAc molecules will be dissociated. This is because as the solution becomes more dilute, the product $[H_3O^+][Ac^-]$ decreases more rapidly than does the [HAc] term. At the same time the concentration of H_2O becomes greater, but because it is so large to start with (about 55.5M), any effect this might have is negligible, which is why no $[H_2O]$ term appears in the equilibrium expression.

For a reaction such as $CH_3COOH(I) + C_2H_5OH(I) \rightarrow CH_3COOC_2H_5(I) + H_2O(I)$ (in which the water concentration does change), dilution will have no effect on the equilibrium; the situation is analogous to the way the pressure dependence of a gas-phase reaction depends on the number of moles of gaseous components on either side of the equation.

Problem Example

The biochemical formation of a disaccharide (double) sugar from two monosaccharides is exemplified by the reaction,

fructose + glucose-6-phosphate → sucrose-6-phosphate

(Sucrose is ordinary table sugar.) To what volume should a solution containing 0.050 mol of each monosaccharide be diluted in order to bring about 5% conversion to sucrose phosphate?

Solution: The initial and final numbers of moles are as follows: td>

	Fructose	Glucose–6–P	Sucrose–6–P
Initial moles:	0.05	0.05	0
Equilibrium moles:	0.0485	0.0485	0.0015

Substituting into the expression for K_c in which the solution volume is the unknown, we have

$$\frac{|\text{suc6P}|}{|\text{Fruc}||\text{gluc6P}|} = \frac{\left(\frac{.0485}{V}\right)}{\left(\frac{.0485}{V}\right)^2} = 0.05$$

Solving for *V* gives a final solution volume of 78 mL.

PHASE DISTRIBUTION EQUILIBRIA

It often happens that two immiscible liquid phases are in contact, one of which contains a solute. How will the solute tend to distribute itself between the two phases? One's first thought might be that some of the solute will migrate from one phase into the other until it is distributed equally between the two phases, since this would correspond to the maximum dispersion (randomness) of the solute.

This, however, does not take into the account the differing solubilities the solute might have in the two liquids; if such a difference does exist, the solute will preferentially migrate into the phase in which it is more soluble.

For a solute S distributed between two phases a and b the process $S_a = S_b$ is defined by the distribution law,

$$K_{ab} = \frac{[S]_a}{[S]_b}$$

in which $K_{a,b}$ is the *distribution ratio* (also called the distribution coefficient) and $[S]_i$ is the solubility of the solute in the phase.

The transport of substances between different phases is of immense importance in such diverse fields as pharmacology and environmental science. For example, if a drug is to pass from the aqueous phase with the stomach into the bloodstream, it must pass through the lipid (oil–like) phase of the epithelial cells that line the digestive tract.

Similarly, a pollutant such as a pesticide residue that is more soluble in oil than in water will be preferentially taken up and retained by marine organism, especially fish, whose bodies contain more oil-like substances; this is basically the mechanism whereby such residues as DDT can undergo *biomagnification* as they become more concentrated at higher levels within the food chain. For this reason, environmental regulations now require that oil-water distribution ratios be established for any new chemical likely to find its way into natural waters. The standard "oil" phase that is almost universally used is octanol, $C_8H_{17}OH$.

In preparative chemistry it is frequently necessary to recover a desired product present in a reaction mixture by extracting it into another liquid in which it is more soluble than the unwanted substances. On the laboratory scale this operation is carried out in a separatory funnel as shown here.

Fig. 8.21

Fig. 8.22

The two immiscible liquids are poured into the funnel through the opening at the top. The funnel is then shaken to bring the two phases into intimate contact, and then set aside to allow the two liquids to separate into layers, which are then separated by allowing the more dense liquid to exit through the stopcock at the bottom. If the distribution ratio is too low to achieve efficient separation in a single step, it can be repeated; there are automated devices that can carry out hundreds of successive extractions, each yielding a product of higher purity. In these applications our goal is to exploit the LeChâtelier principle by repeatedly upsetting the phase distribution equilibrium that would result if two phases were to remain in permanent contact.

Problem Example

The distribution ratio for iodine between water and carbon disulfide is 650. Calculate the concentration of I_2 remaining in the aqueous phase after 50.0 mL of 0.10M I_2 in water is shaken with 10.0 mL of CS_2.

Solution:

The equilibrium constant is,

$$K_d = \frac{C_{CS_2}}{C_{H_2O}} = 650$$

Let m_1 and m_2 represent the numbers of millimoles of solute in the water and CS_2 layers, respectively. K_d can then be written as $(m_2/10 \text{ mL}) \div (m_1/50 \text{ mL}) = 650$. The number of moles of solute is $(50 \text{ mL}) \times (0.10 \text{ mmol mL}^{-1}) = 5.00$ mmol, and mass conservation requires that $m_1 + m_2 = 5.00$ mmol, so $m_2 = (5.00-m_1)$ mmol and we now have only the single unknown m_1. The equilibrium constant then becomes

$((5.00-m_1)$ mmol/ 10 mL) ÷ (m_1 mmol/ 50 mL) = 650.

Simplifying and solving for m_1 yields $(0.50-0.1)m_1/(0.02\ m_1) = 650$, with$m_1 = 0.0382$ mmol. The concentration of solute in the water layer is (0.0382 mmol)/ (50 mL) = 0.000763 M, showing that almost all of the iodine has moved into the CS_2 layer.

The six Problem Examples presented above were carefully selected to span the range of problem types that students enrolled in first–year college chemistry courses are expected to be able to deal with. If you are able to reproduce these solutions on your own, you should be well prepared on this topic.

The first step in the solution of all but the simplest equilibrium problems is to sketch out a table showing for each component the *initial* concentration or pressure, the *change* in this quantity (for example, +2*x*), and the *equilibrium*values (for example.0036 + 2*x*). In doing so, the sequence of calculations required to get to the answer usually becomes apparent.

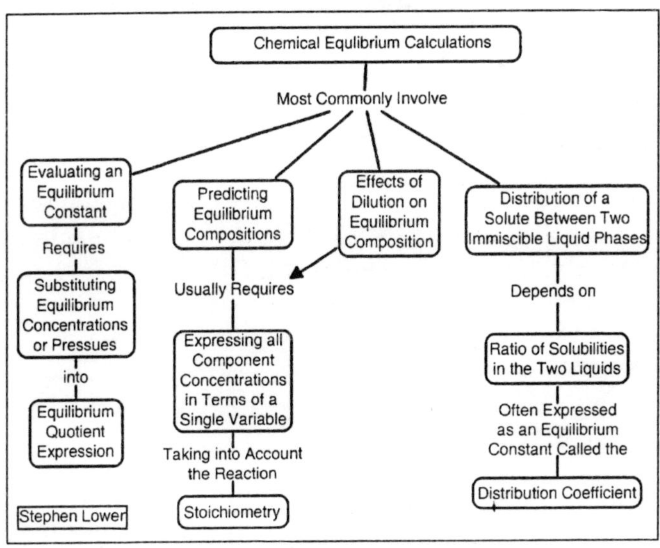

Fig. 8.23

Equilibrium calculations often involve quadratic–or higher-order equations. Because concentrations, pressures, and equilibrium constants are seldom known to a precision of more than a few significant figures, there is no need to seek exact solutions. Iterative approximations are adequate and convenient.

Phase distribution equilibria play an important role in chemical separation processes on both laboratory and industrial scales. They are also involved in the movement of chemicals between different parts of the environment, and in the bioconcentration of pollutants in the food chain.

Chapter 9

The Solid and Liquid Phases

SPECIFIC CONCEPTS

The three common phases of matter are gas, solid, liquid:

1. The structures of many solids can be understood in terms of a few atom arrangements.
2. The liquid phase shows properties intermediate between those of solid and gas.
3. All liquids tend to evaporate and establish a vapour pressure in a closed container.

The vast majority of known substances are solids under ordinary conditions of pressure and temperature. The purpose of this chapter is to provide an understanding of the structures of crystalline solids at the atomic/molecular level. Be forewarned that a study of such structures requires a sound basis in geometry and an ability to visualize in 3–dimensions.

Further, it is impossible to obtain an adequate understanding of the important structural ideas from text and 2-dimensional figures. Work with 3-dimensional models is essential. Read through the chapter, understand it as best you can, then refine your understanding with models in the laboratory. The ideas discussed in this chapter provide a solid base for the understanding of two major classes of materials: metals, in which all atoms in the structure are the same; and ionic compounds, composed of cations of one substance and anions of another.

MACROSCOPIC PROPERTIES OF SOLIDS

Solids have a number of observable properties in common:

- They are rigid, maintaining fixed shape and volume independent of container.
- The molar volumes of solids are small.
- Solids are incompressible.
- Solids do not flow, and diffusion in solids is extremely slow.
- Many solids occur as crystals—well-defined shapes with structures suggesting an orderly internal arrangement of atoms, molecules, or ions.

To rationalize these properties we need a simple model of the solid phase. In designing such a model, we make use of two facts:

1. Since the molar volume of a solid is much less than the molar volume of the gas phase of a substance, the molecules must be crowded together in the solid;
2. The orderly shapes of crystals suggests an orderly arrangement of particles.

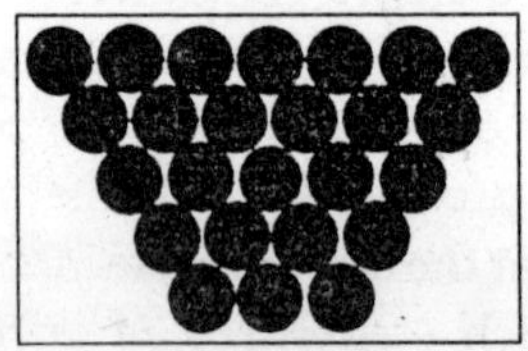

Fig. 9.1

A simple model, then, is that solids consist of tightly–packed orderly arrangements of particles. Figure 9.1 shows a 2–dimensional representation of a solid. Note first how close together the particles are compared with the gas phase. This crowding allows very little free volume to the molecules, which strongly resist further compression. The solid is incompressible. It is also clear why diffusion is slow in solids, because each particle is locked in place by the surrounding ones. Only at the surface of the solid, where a particle has neighbours on only 3 sides, can diffusion occur. Although it is not conveyed by Figure, the particles are in constant motion, just as in the gas phase. A molecule is constantly jiggling in place, bumping into neighbours that hold it in a cage.

Just as for gases, the kinetic energies of the particles (atoms, molecules, or ions) of the solid are distributed according to the Maxwell–Boltzmann distribution, with the kinetic energy of an average particle having the value 3kT/2. The major difference between the motion in gas and solid is that in the solid, a particle is hemmed in and undergoes many more collisions per unit time than in the gas.

The movement of a particle in the solid is much like a vibration, because the molecule stays in the same location. If the temperature of the solid is increased, the particles vibrate faster, but are still not free to move about because intermolecular forces from the neighbours hold them in place.

Finally, the arrangement of particles is very orderly, not at all chaotic like the gas. Each particle is surrounded by a perfect hexagon of other particles; and particles are arranged in rows that remain straight over long distances. It is easy to imagine that this orderly arrangement at the molecular level is manifested in beautiful crystalline shapes at the macroscopic level.We now begin to look in detail at the orderly arrangements of particles in solids.

The Lattice Concept

We will describe structures of crystalline solids in terms of the lattice. A lattice is a regular, repeating pattern of points in space. Wallpaper patterns are examples of 2–dimensional lattices. A wallpaper pattern and the equivalent lattice are shown in Figure 9.2.

The lattice is more fundamental than the particular wallpaper pattern shown, because wallpaper patterns using many different design units could be generated from the same lattice. The key feature of the lattice is the arrangement of points. A small collection of lattice points that contains all of the spatial information in the lattice pattern is called a unit cell. A unit cell is shown outlined in the lattice in Figure. The unit cell represents the lattice, because the whole lattice can be reproduced from the unit cell by moving (translating) the unit cell in directions parallel to its sides by distances equal to the lengths of its sides.

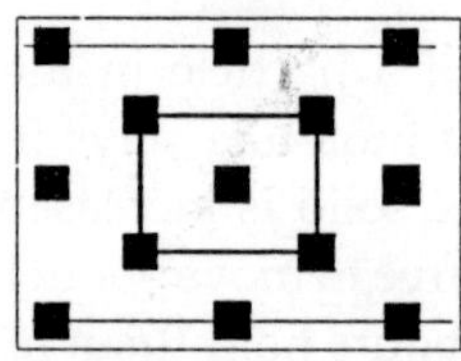

Fig. 9.2 Shows a Second Small Collection of Points That Might be Considered as a Unit Cell.

Movement of this collection in directions parallel to its sides does indeed reproduce the entire lattice. However, the unit cell in Figure 9.2 is simpler than that in Figure 9.3 because it involves fewer points—one point at each of its 4 corners. A unit cell having lattice points at its corners only is called a primitive unit cell. The unit cell in Figure 9.3 is a non-primitive unit cell because it has an additional point at its center.

If it is possible to represent a lattice with a primitive unit cell, we will do so. Crystalline solids are represented by 3-dimensional lattices, one example of which is the simple cubic lattice. The unit cell of the simple cubic lattice is shown in Figure 9.4. This is a primitive unit cell because it has points only at the corners. Movement of the unit cell by the distance, d, in directions parallel to its 3 dimensions reproduces the entire simple cubic lattice. The simple cubic lattice is analogous to an orderly arrangement of building blocks stacked face to face in 3-dimensions.

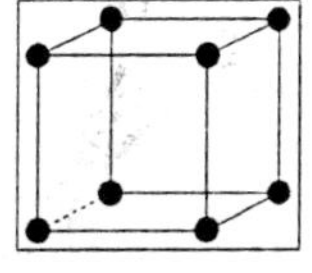

Fig. 9.3

It was shown in the 1800s by Bravais that there are only 7 shapes (called crystal types) that a unit cell can have and only 14 possible distributions of points (lattice types) within these 7 shapes. The 7 shapes, characterized by the relative magnitudes of the repeat distances in three directions and the angles between these directions, are summarized in Table.

Table: The Bravais Crystal Types

System	Edge Lengths	Angles
Cubic	a = b = c	all 90°
Tetragonal	a = b not equal to c	all 90°
Orthorhombic	a not equal to b not equal to c	all 90°
Monoclinic (once-angled)	a not equal to b not equal to c	$\alpha = \beta = 90°$, γ not equal to 90°
Triclinic (thrice-angled)	a not equal to b not equal to c	α not equal to β not equal to γ not equal to 90°
Rhombohedral	a = b = c	= 90°
Hexagonal	a = b not equal to c	$\alpha = \beta = 90°$; $\gamma = 120°$

We will discuss cubic and hexagonal lattices in the context of a discussion of close-packing ideas.

Close Packing

What is the most efficient way of arranging spheres in 3–dimensional space? We take "efficient" to mean economical in terms of space. We begin by building one layer of spheres that are arranged as closely to one another as possible. Spheres are lined up in straight rows, with the spheres in one row fitting snugly into the indentations in adjacent rows. Each sphere is surrounded by a perfect hexagon of spheres, all touching it. There are numerous gaps, or holes, in the layer, of triangular shape. For descriptive purposes here, we label these holes "u" if the triangle has a single vertex pointing up, or "d" if the triangle has vertex down.

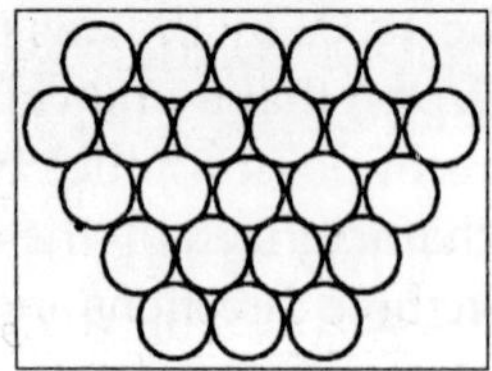

Fig. 9.4

We now add a single sphere of a second layer on top of the first (the black–shaded sphere in Figure 9.4). The added sphere rests in a low point of the first layer, corresponding with a triangular hole, in contact with 3 spheres of the first layer. The 4 touching spheres lie at the vertices of a regular tetrahedron. There remains some unfilled space at the center of this 4–sphere group that is called a tetrahedral hole. Each time we add a sphere to the second layer, we create a tetrahedral hole between the layers. When additional spheres are added to the second layer, it turns out that they must all go in either "u" holes or "d" holes, but not both. Thus only half of the triangular depressions in the first layer can be used for placement of 2nd-layer spheres.

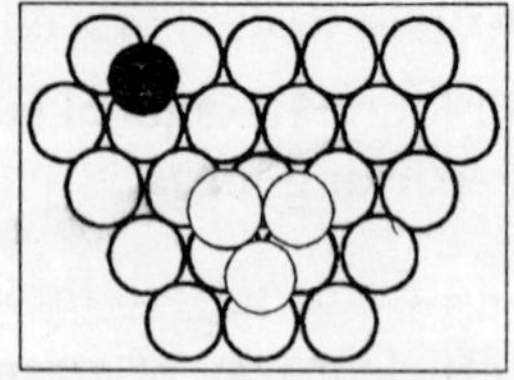

Fig. 9.5

Triangular holes in the 2nd layer lie directly over the unused holes of the first layer, so that channels pass right through both layers. The point at the center of such a channel is surrounded by 6 spheres (gray–shaded in Figure 9.5), arranged as 2 equilateral triangles, one triangle in the 1st and one in the 2nd layer. These 6 spheres form a regular octahedron, a geometrical solid with 6 vertices and 8 equilateral triangular faces. A regular octahedron is shown in Figure 9.6. The space surrounded by the 6 spheres is called an octahedral hole. An octahedral hole is larger than a tetrahedral hole because 6 spheres cannot be crowded as closely together as 4 spheres.

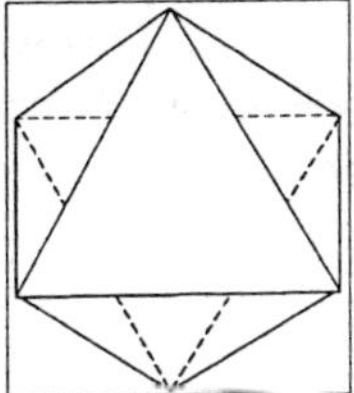

Fig. 9.6

There are 2 ways in which a third layer of spheres can be added to the existing 2 layers in close-packed fashion. In the first way, spheres are place directly over spheres of the first layer. The third and first layers then have spheres in the same locations. The 4th layer then goes over the second, the 5th over the 3rd, and so on, so a pattern that repeats every other layer is obtained. This is characterized as a 1212... pattern. Figure 9.7 shows the unit cell of this 3–dimensional lattice. The unit cell is hexagonal in shape; the 1212... close–packing arrangement is therefore called hexagonal close packing (hcp). Note that the coordination number (the number of nearest neighbours) of any sphere in hcp is 12.

Fig. 9.7

The second way of adding a third layer is to place spheres directly over the octahedral holes between the first two layers. In this case, the third-layer spheres do not lie over spheres in layer 1. This is still a close-packing arrangement because the coordination number of any sphere is 12, just as in hcp. The 4th layer of spheres is then placed directly over the spheres in the first layer. This type of close packing is designated 123123... to indicate that the pattern repeats after 3 layers. The unit cell for 123... close packing is shown in Figure 9.8. Although it may not be obvious from the figure 9.7, this unit cell is a cube (this is a situation in which a 2–dimensional drawing does not adequately convey what is true). This type of close packing is therefore called cubic close packing (ccp).

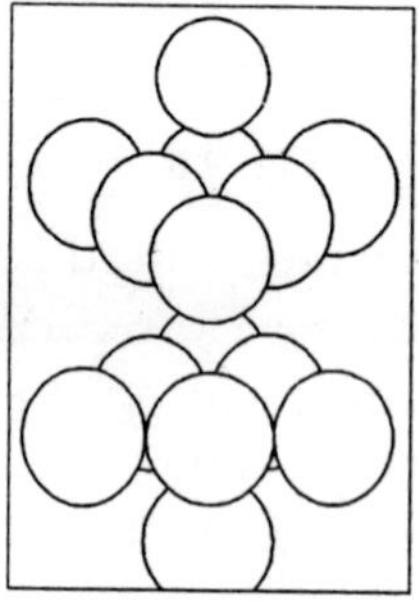

Fig. 9.8

Hcp and ccp are the two ways of close packing spheres in 3 dimensions.

They have several features in common:

1. Spheres occupy 74% of the available space;
2. Each sphere has 12 nearest neighbours that touch it (coordination number);
3. The number of octahedral holes in the lattice is exactly the same as the number of spheres in the lattice, and there are twice as many tetrahedral holes as spheres! This is not obvious and will be addressed again later.

There are a number of substances that crystallize in one or the other of these two packing arrangements. Beryllium, magnesium, and zinc metal all form hexagonal close packed lattices. Nickel, copper, gold, and aluminum form ccp lattices. Cubic close packing is particularly common. We will take a closer look at cubic lattices now.

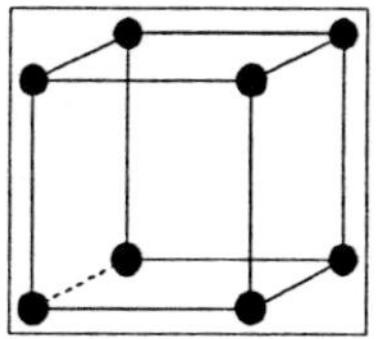

Fig. 9.9

The three Bravais cubic lattices are shown in Figure 9.9. The simple cubic (sc) lattice is primitive, with points (spheres or atoms) only at the vertices of the cube. Metallic polonium and non–metallic dioxygen (O_2) crystallize with the simple cubic lattice. The coordination number of an atom in the simple cubic lattice is 6. To determine this, pick one atom in the unit cell; find an atom that appears to be closest to the atom you picked; then count the number of other atoms that are at this same distance from the atom you picked. In counting coordination numbers, you may have to go outside of the single unit cell pictured in Figure.

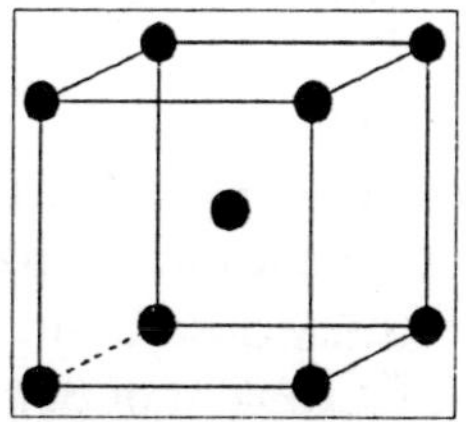

Fig. 9.10

In the simple cubic lattice, identical unit cells surround the pictured one on all sides. Some of the nearest neighbours of your picked atom lie in these neighboring unit cells. We say that the simple cube has 1 atom per unit cell (1 atom/uc). The body-centered cubic (bcc) lattice is non-primitive, with points at the corners and the center of the unit cell (the center of a cube is called the cube body center). Metallic iron and barium crystallize in the bcc lattice. The coordination number is 8 for this lattice (count the nearest neighbours of the body–center atom), and the body-centered cube has 2 atoms per unit cell. Finally, the face–centered cubic (fcc) lattice is also non-primitive with points at each corner and in the center of each of the 6 faces of the cube. Metallic nickel and copper adopt this structure. The coordination number is 12 for this lattice, which means that it must be close packed!

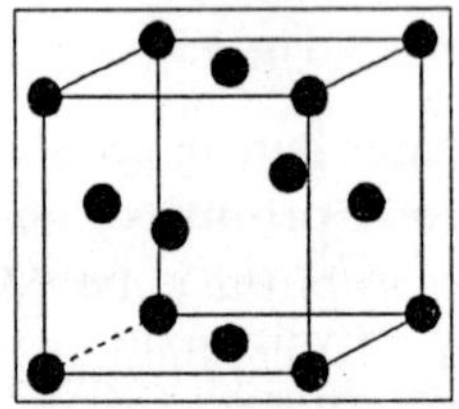

Fig. 9.11

A body diagonal runs on a straight line from one corner, through the body center of the cube, to the diagonally opposite corner. As you look along the body diagonal, you will see the pattern of spheres shown in Figure 9.11: a single sphere; a triangle consisting of 6 spheres; a second triangle of 6 spheres, but rotated 60° with respect to the first one; and finally another single sphere immediately below the first one. Figure 9.11 is an attempt to outline the face-centered cube in a ccp arrangement of spheres. The cube is admittedly difficult to visualize from the drawings. The face-centered cubic lattice has 4 atoms per unit cell.

We now address the problem of determining the number of atoms per unit cell. This is quite simple once it is realised that atoms are generally shared among several unit cells and therefore must be partitioned accordingly. An atom at a corner (vertex) of a cube is shared among 8 unit cells. To see this, imagine the simple cube in Figure 9.11 to be a building block, and focus on the upper right front corner of the building block. Imagine that this corner is labelled "a".

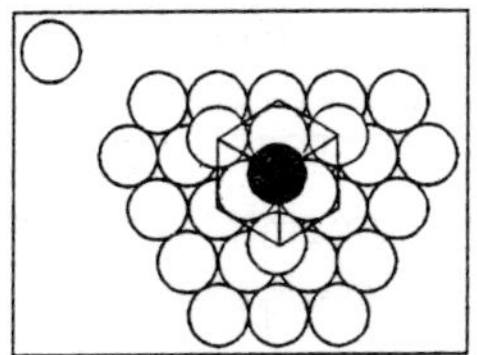

Fig. 9.12

In your mind, place identical blocks on top, to the right, and in front of the block in the figure 9.12. Then place additional blocks in the remaining places adjoining these blocks to generate the 8–block arrangement in Figure 9.13. Point "a" is now at the center of this collection, and is simultaneously a corner for each of the 8 blocks; 8 blocks share a vertex. This point cannot be said to belong entirely to any particular block. Instead, each block gets 1/8 of it.

The counting rule for a corner atom, then, is that only 1/8 of it is assigned to a single unit cell. Similar thinking leads to the conclusion that a face-center atom is shared between 2 unit cells, so 1/2 of it is assigned to a particular unit cell; and a body–center atom is not shared at all, so it belongs entirely to the unit cell that it is found in. The atom count for a body–centered unit cell then goes like this:

Atoms/uc = 8 corner atoms × 1/8 + 1 body-center atom × 1

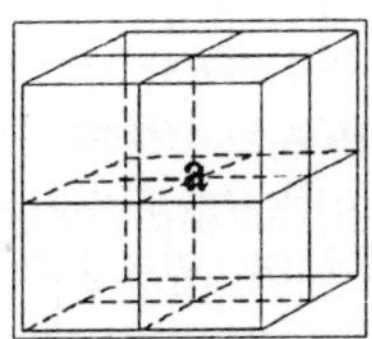

Fig. 9.13

We will have occasion also to partition points located along edges of the cube. Edges are the lines connecting adjacent vertices. Since an edge is common to 4 cubes arranged in a square, 1/4 of any point along an edge is assigned to a particular cube. The point-counting rules are summarized in Table.

Table. Atom Counting Rules for Cubic Unit Cells

Atom location	Fraction Assigned to uc
Corner	1/8
Edge	1/4
Face	1/2
Body-center	1

If, as we have said, the face-centered cubic lattice is in fact cubic close packed, then there should be tetrahedral and octahedral holes in the fcc lattice. To locate these holes it is convenient to place our cube in a coordinate system. The origin of coordinates is located at the lower left front corner of the cube, with the x–axis along the bottom front edge, the y–axis along the left vertical edge, and the z axis along the left bottom edge. This is shown in Figure 9.14.

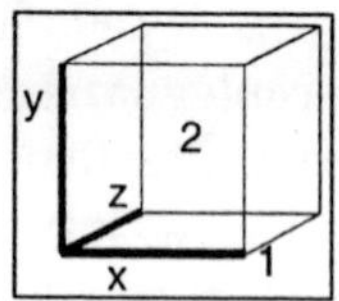

Fig. 9.14

We specify distances along the 3 axes in terms of fractions of the cube edge length. The set of coordinates (1,0,0) means to proceed 1 edge length along x. These are the coordinates of the lower right front vertex of the cube. Similarly, (1/2,1/2,1/2) specifies the body center of the cube, labelled 2 in the figure 9.15.

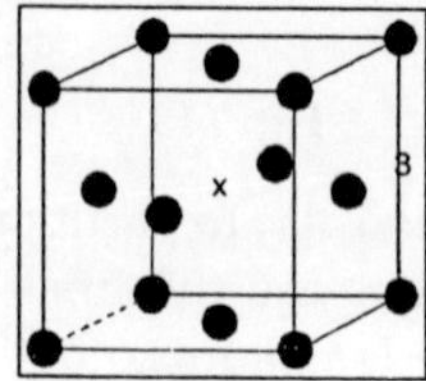

Fig. 9.15

We find the most obvious octahedral hole at the body

center. This point is surrounded by 6 atoms all at the same distance from it. These are the atoms at the face centers of the cube. Similar octahedral holes are located at the cube edge centers. For example, the point labelled 3 in Figure 9.15 is an octahedral hole. It is surrounded at equal distance by the atoms at (1/2,1/2,1), (1,0,1), (1,1,1), and (1,1/2,1/2). The remaining 2 atoms are not part of the unit cell shown, but can still be located with coordinate sets (1 1/2,1/2,1) and (1,1/2,–1 1/2). (Translating these coordinates is good practice in the use of the unit cell coordinate system.) The cube has 12 edges, so there are 12 holes of this type. The counting rules tell us that only 1/4 of each of these edge-centered octahedral holes belongs to the unit cell of interest. The total number of octahedral holes in the unit cell is then 1(body center) + 12(1/4) = 4. The number of octahedral holes is the same as the number of atoms in the unit cell!

It is defined by the four shaded atoms, and has coordinates (1/4,1/4,1/4). There are 7 other tetrahedral holes in the unit cell, all with coordinate combinations involving 1/4 and 3/4, and all defined by 1 corner and 3 face-center atoms. Each tetrahedral hole lies entirely within the cube body, so none of them are shared with other unit cells. The number of tetrahedral holes per fcc unit cell is therefore 8, twice the number of atoms in the unit cell.

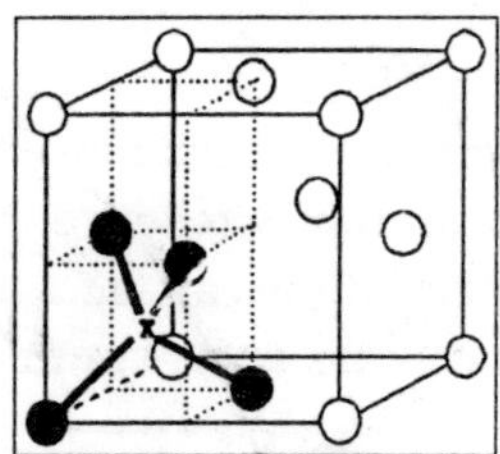

Fig. 9.16

Octahedral and tetrahedral holes are important in the discussion of ionic compounds.

IONIC COMPOUNDS

The ideas discussed thus far apply to crystals in which the units—spheres or atoms—are all identical. Crystals of pure metals, with close-packed metal atoms, and even crystals of molecular solids (for example, methane, CH_4, in which there is a methane molecule at each lattice point) are examples of real situations in which these ideas apply. Ionic compounds are different because there are two distinctly different units—the cation and the anion—that must be accommodated in the lattice.

The resulting lattices are often called compound lattices because they contain two different components. The cation and the anion differ in charge, of course, and in chemical identity (usually). They also differ in size. Generally speaking, cations are smaller than anions, for easily-understood reasons. Cations contain an excess of protons over electrons. The Coulomb pull of the excess positive charge causes the electron cloud to contract, so that it is smaller than in the corresponding atom (Na^+ is smaller than Na). Anions contain an excess of electrons over protons.

The repulsion between the electrons exceeds that in the corresponding neutral atom, causing an expansion of the electron cloud (Cl^- is larger than Cl). Except for a few rare cases, the cation in an ionic compound is smaller than the anion. In forming a lattice, the space requirements are more important for the larger anion, which tends to dictate which lattice is adopted. The (smaller) cations then fit into the holes, or interstices, of the anion lattice. Many common ionic compounds have lattices that consist of cubic close packed anions, with cations in tetrahedral and/or octahedral holes. Stoichiometries of up to 3 cations per anion can be accommodated in this framework. We now discuss some common ionic lattices.

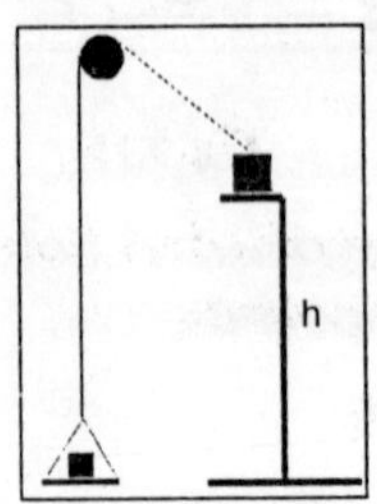

Fig. 9.17

The Sodium Chloride (Rock Salt) Lattice

The NaCl lattice is shown in Figure 9.18. It can be described as a cubic close packed (face-centered cubic) anion lattice, with Na^+ ions in the octahedral holes. The coordination number (number of nearest neighbours of opposite charge) of each Na^+ ion is 6, and of each Cl^- ion is 6. The cation and anion must have equal coordination numbers because the stoichiometry of the compound is 1:1 (1 cation per anion).

The unit cell contains 4 formula units of NaCl (4 Cl^- ions, which define the fcc unit cell; 4 Na^+ ions because there are 4 octahedral holes per unit cell). Many ionic compounds exhibit the NaCl lattice, including most of the group 1 halides, most of the group 2 oxides, and a number of d–metal oxides.

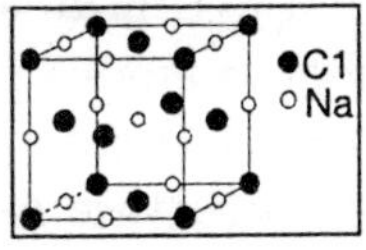

Fig. 9.18

The Zinc Blende Lattice

Zinc blende is the common name for one crystal modification of the common mineral, ZnS. The other form is known as wurtzite. Wurtzite consists of an hcp array of S^{2-} ions with Zn^{2+} ions in one-half of the tetrahedral holes; it will not concern us further here, because we are restricting our attention to cubic lattices. Zinc blende consists of a ccp array of S^{2-}, with Zn^{2+} in half of the tetrahedral holes.

Since there are 8 tetrahedral holes in the sulfide ion lattice, but only 4 S^{2-} ions, only half of the holes are used to accommodate 4 Zn^{2+} ions. This is consistent with the 1:1 stoichiometry of the compound. The zinc ions occupy tetrahedral holes so that they can be as far apart from one another as possible to minimize repulsions between positive ions. If the first Zn^{2+} ion occupies the hole at (1/4,3/4,1/4), the other three will be in holes at (3/4,1/4,1/4), (1/4,1/4,3/4), and (3/

4,3/4,3/4). You should locate these holes and verify that the Zn^{2+} ions describe a tetrahedron. The cation and anion coordination numbers are both 4 in zincblende. As for NaCl, there are 4 formula units per unit cell. Other ionic compounds that adopt the zinc blende lattice include BeS, CdS, HgS, and AgI.

The Calcium Fluoride (fluorite) Lattice

The Ca^{2+} ion is virtually the same size as the F– ion, one of those rare situations referred to above, and forms a face–centered cubic lattice. The F^- ions fill all of the tetrahedral holes in the cation lattice. Since there are 8 such holes per 4 Ca^{2+} ions, the stoichiometry is nicely accommodated.

The coordination number of the Ca^{2+} ion is 8, and that of the F^- ion is 4. (Note that the number of cations per formula unit multiplied by the cation coordination number is equal to the product of the number of anions per formula unit and the anion coordination number.) There are again 4 formula units per unit cell. The fluorite structure is very common for ionic compounds of 1:2 (or 2:1) stoichiometry.

The Cryolite Lattice

Cryolite is the common name for the mineral, Na_3AlF_6. Though rare, this substance is of extreme commercial value, because it is used in the process of extraction of aluminum from bauxite ore. Cryolite consists of Na^+ cations and AlF_6^{3-} anions. It adopts a lattice in which the anions are cubic close packed (fcc) and the cations occupy all of the octahedral and tetrahedral holes. The Diamond Lattice. Diamond is one crystal modification of the element carbon. It is, of course, not ionic, nor even a compound. It is dealt with here because it, too, is based on a fcc unit cell. The diamond lattice is just like the zinc blende lattice, with carbon atoms in place of both Zn^{2+} and S^{2-} ions. We therefore describe it as a fcc array of carbon atoms with carbon atoms in one-half the tetrahedral holes.

The Cesium Chloride Lattice

It is a more open lattice than those considered thus far. This lattice consists of a simple cubic array of chloride ions, with Cs^+ ions in cubic holes. A cubic hole is the space at the body center of a simple cubic unit cell.

There is 1 formula unit of CsCl per unit cell, and the coordination number of both cation and anion is 8. Ionic compounds that adopt the CsCl lattice include CsBr, CsI, the halides of thallium(I), and NH_4Cl.

Radius Ratio Rules. Compounds of 1:1 stoichiometry commonly adopt one of three lattices. Of interest now is whether or not we can predict, or at least understand, why a particular compound favors one lattice over the others.

A number of factors are important in choice of lattice, but one of these dominates in importance: the relative sizes of cation and anion. Because we view the cation as occupying holes in the anion lattice, we begin by calculating the sizes of the various hole types in relation to the size of the anion involved. The most straightforward calculation is for the octahedral hole, which is shown in Figure.

The figure shows one face of the fcc unit cell, with the anions just touching along a face diagonal of the cube. There is an octahedral hole at the center of each of the 4 cube edges shown; we focus on the right edge. The size of the octahedral hole will be taken as the distance from the edge center to, say, the top right anion, along the cube edge. (The distances from the hole to the bottom right anion and to the face-centered anion are the same.) We can express the right edge length as a sum of anion and hole radii as in equation 7–4–1:

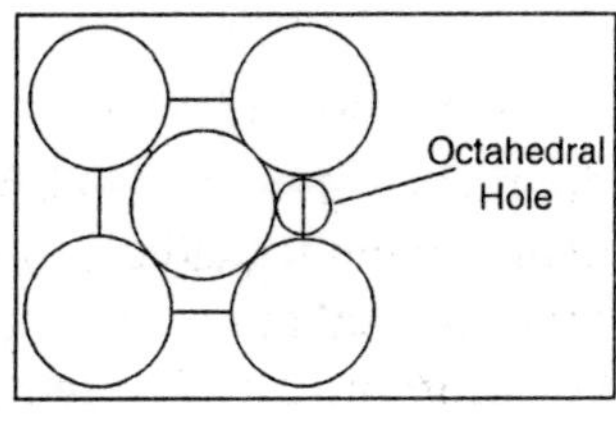

Fig. 9.19

$$(7\text{–}4\text{–}1): a = 2r^- + 2r^+$$

Here r^- is the anion radius, r^+ the hole radius. (The + designation is used in anticipation of putting a cation in the hole.) Similarly, we express the length of the face diagonal (fd) in terms of anion radii. This is easy, since the anions touch along this line:

$$(7\text{–}4\text{–}2): fd = 4r^-$$

Finally, we relate the fd and edge length via the Pythagorean Theorem:

$$(7\text{–}4\text{–}3): (fd)^2 = 2a^2$$

Take the square root of both sides of 7-4-3:

$$(7\text{–}4\text{–}4): fd = a(2)^{1/2}$$

Substitute 7–4–4 in 7–4–2 and rearrange for a:

$$(7\text{–}4\text{–}5): a = 2(2)^{1/2}\, r^-$$

Substitute for a in 7-4-1 *and solve for* r^+:

$$(7\text{–}4\text{–}6): r^+ = ((2)^{1/2}-1)\, r^- = 0.414\, r^-$$

The octahedral hole is somewhat less than half as large as the anions that create the hole. By similar geometric methods, equations 7–4–7 and 7–4–8, for tetrahedral and cubic holes respectively, can be developed:

$$(7\text{–}4\text{–}7): r^+ = 0.224\, r^-$$

$$(7\text{–}4\text{–}8): r^+ = 0.732\, r^-$$

You should try to obtain these expressions, making use of right triangles and Pythagorean relationships. (In addition to equation 7–4–4, which relates the face diagonal and edge length of a cube, equation 7–4–9 relates the cube body diagonal (bd) to the edge length:

$$(7\text{–}4\text{–}9): bd = (3)^{1/2}\, a$$

According to Equations 7–4–6 to 7–4–8, a tetrahedral hole is about 1/2 the size of an octahedral hole and about 1/3 the size of a cubic hole. Only cations that are relatively small compared with the anion can be accommodated in tetrahedral holes. As cation size increases relative to anion size, the cation will prefer a larger hole, with a larger coordination number of anions.

Equations 7–4–6 to 7–4–8 are the basis for the following radius-ratio rules:

- If the ratio of the cation and anion radii (r^+/r^-) lies in the range between 0.224 (equation 7–4–7) and 0.414 (equation 7–4–6), the cation prefers to occupy a tetrahedral hole.
- If r^+/r^- lies in the range between 0.414 and 0.732, the cation prefers an octahedral hole.
- If r^+/r^- exceeds 0.732, the cation prefers a cubic hole.

These rules make good sense. If $r^+/r^- < 0.224$, the cation is too small for the tetrahedral hole. The anions will be touching, but will not be touching cations. Since it is the cation-anion attractions that stabilize the lattice, whereas anion-anion or cation-cation repulsions tend to destabilize it, this is not a favorable situation.

When $r^+/r^- = 0.224$, the cation just fits the tetrahedral hole. Although anions are still touching, they now touch cations as well. This is a stable arrangement. As r^+/r^- becomes > 0.224, the cation is too big for the tetrahedral hole and starts to push the anions apart.

The cation-anion attractions are maintained, because cations and anions still touch, but now anion-anion repulsions are decreased because the anions no longer touch. This is an even better arrangement than the perfect fit. This situation continues to improve until r^+/r^- becomes equal to 0.414.

At this value, the cation prefers to occupy an octahedral hole because it can increase its coordination number without increasing anion-anion repulsion to any significant extent.

The number of cation-anion attractive interactions increases from 4 to 6, giving a more stable lattice. As r^+/r^- increases between 0.414 and 0.732, cations in octahedral holes maintain their contact with 6 anions while pushing the anions further and further apart, decreasing repulsions between them.

Throughout this range of r^+/r^-, occupation of an octahedral hole becomes increasingly favorable until the radius ratio becomes 0.732. The cation can now increase its coordination number to 8 anions, again increasing the number of attractive cation-anion interactions. The radius ratio rules are summarized in Table. Radii for common cations and anions are given in Table.

Table: Radius Ratio Rules

Range of r^+/r^-	Lattice Type	Example
0.224–0.414	ZnS	ZnS
0.414–0.732	NaCl	KCl
> 0.732	CsCl	CsCl

Ion radii and the radius ratio rules can be used to predict the probable choice of lattice for an ionic compound.

Example. The radii of the Na^+ ion and the Br^- ion are 116 and 182 pm, respectively. Which of the three 1:1 lattices will be adopted by NaBr?

Solution. Calculate r^+/r^-:

$$r^+/r^- = 116/182 = 0.64$$

This falls between 0.414 and 0.732. The cation will prefer the octahedral hole. NaBr should therefore have the rock salt structure. It does. The radius ratio rules provide useful guidelines to the lattice structures of ionic compounds. It is important to realise, however, that they are not hard and fast; the predictions made using the rules are often wrong. For example, in the zincblende form of ZnS, Zn^{2+} ions occupy tetrahedral holes in a fcc lattice of S^{2-} ions. In fact, this compound is the prototype for this type of structure.

However, the radius ratio rules predict that the Zn^{2+} ions should occupy octahedral holes! The radius ratio rules fail frequently because the relative size of the ions is not the only factor that is important in determining structure. Of particular importance is the tendency towards at least some covalent bonding between cation and anion. Covalency involves directional requirements, as we have seen in Chapter 3, and may influence the type of hole preferred by the cation.

7-5 X-RAY CRYSTALLOGRAPHY

In 1912, von Laue predicted that crystals could serve as

diffraction gratings if the wavelength of radiation used were approximately the same as the distance between planes of atoms in the crystals.

This spacing was known to be on the order of $1 * 10^{-10}$ m, corresponding to wavelengths in the x–ray region of the electromagnetic spectrum. Subsequently, WH Bragg discovered that when X–rays of appropriate wavelength are directed onto the smooth face of a crystal, the X–rays are diffracted and a pattern characteristic of the lattice structure of the crystal is obtained.

This observation is the foundation for the modern technique of X–ray Crystallography, the most powerful method that we have for determining molecular structure. Like the types of spectroscopy discussed in x–ray crystallography is based on the interaction of light with matter. Unlike spectroscopy, however, the light is not absorbed by the matter in this case. Instead, it is scattered from the atoms of the crystal in a very definite way.

Analysis of the scattering enables the arrangements of atoms within the crystal to be deduced. The analysis of x–ray diffraction patterns is complex, even for materials of simple structure. However, thanks to the efforts of many crystallographers over many years, it is today done quite quickly and routinely. Once a suitable crystal of a substance is grown, the modern crystallographer is able to mount it in a fully automated x–ray diffractometer, the instrument that is used for the x–ray diffraction study.

The diffractometer automatically obtains all of the necessary data for the determination of the lattice structure. The data are then fed to one of a number of computer programme that have been developed and refined for the analysis of such data, and the structure is "solved." The output of the programme reveals not only the lattice structure, but also the locations of all of the atoms of a molecule of the substance. Structures of substances ranging from simple salts or molecular substances to huge complex proteins and enzymes can be determined using this approach. The structures of two molecules, obtained from x–ray diffraction data, are shown in Figure Figure shows the arrangement of atoms in the relatively simple molecule, phthalocyanine. Figure 9.20 shows part of the structure of the protein, hemoglobin, which has the function of carrying oxygen from the lungs to the tissues in living organisms.

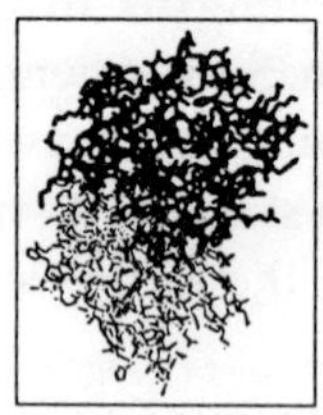

Fig. 9.20

We now look at a couple of examples of simple applications of crystallographic data.

Example. Aluminum crystallizes in a cubic lattice. X–ray diffraction reveals that the edge length of the unit cell is 405.0 pm. The density of Al is 2.7 g/cm^3. Which cubic lattice does Al adopt?

Solution. We adopt the following strategy:

Unit cell mass → Number Al atoms per uc → Lattice type

We can obtain the unit cell mass from the density by calculating unit cell volume from the cube edge length.

Volume of uc = $a^3 = (4.05 * 10^{-8} \text{ cm})^3$

$= 6.643 * 10^{-23} \text{ cm}^3/\text{uc}$

Mass of unit cell = r × Volume uc

$= 2.7 \text{ g/cm}^3 * 6.643 * 10^{-23} \text{ cm}^3/\text{uc}$

$= 1.794 * 10^{-22}$ g

Moles Al per uc = mass/molar mass

$= 6.647 * 10^{-24}$ moles Al

Atoms Al per uc = $6.647 * 10^{-24}$ moles * $6.022 * 10^{23}$ atoms mol

= 4.0

The unit cell is face-centered cubic. This example shows the connection between macroscopic and microscopic properties. Note that in a reversal of the above procedure, knowledge of the type of cubic lattice could be used to experimentally measure Avogadro's number, N_o. Example. In early nuclear reactors, fuel rods were made of uranium metal, with a cubic close packed lattice. These suffered from the problem that the fuel rods frequently burst out of their protective sheaths, which could obviously cause major safety problems. Why did this occur, and what could be done to prevent it?

Solution. Nuclear fission (the spontaneous breakup of a radioactive nucleus) produces 2 nuclei for each one that decays. One possible decay route is shown below:

$$^{235}U \rightarrow {}^{90}Sr + {}^{143}Xe + 2\ {}^{1}n$$

In the ccp U lattice, the holes are too small to accommodate these extra nuclei. The resulting expansion of the lattice caused bursting of the protective sheaths. The solution of the problem was to make the fuel rods from UO_2 (yellow cake), which consists of ccp U atoms with O atoms in the tetrahedral sites. The larger octahedral holes in this lattice were able to accommodate the fission fragments without expansion.

THE BRAGG EQUATION

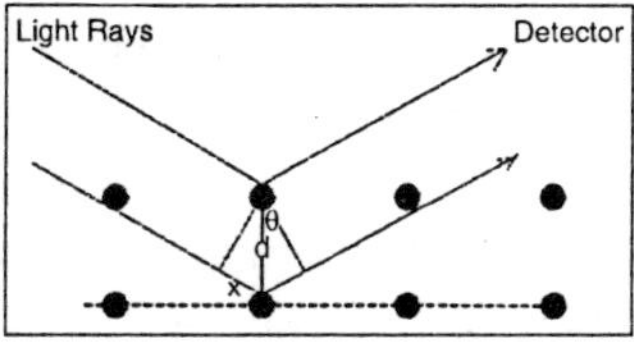

Fig. 9.21

Figure 9.21 shows two planes of atoms in a crystal. A source of X–rays at the left emits radiation, two beams of which are shown impinging on the planes of atoms and scattering off to the right, where they are "seen" at a detector.

The angles made by the incident beam and the deflected beam with the planes of atoms are the same. A diffraction pattern is a series of alternating light and dark lines (sometimes spots), as shown in Figure 9.21.

The white spots result from the arrival at the detector of reflected beams that are all in-phase. That is, their waves are matched up peak for peak and trough for trough. Waves that are in phase add together to give a wave with larger peaks and larger troughs—*i.e.*, more intensity. This is called constructive interference.

The white spots in the diffraction pattern correspond to deflection angles that give in-phase waves. Similarly, the dark areas correspond to angles that give out–of–phase waves. Here the peak of one wave arrives at the detector at the same time as a trough of a second wave arrives. Such waves cancel each other, and very little or no signal is seen at the detector. This is called destructive interference.

The criterion that must be met in order that two waves arrive at the detector in phase is that the difference in the distances travelled by them must correspond to an integer number of wavelengths. This will give the maximum intensity at the detector. If the difference in distance travelled corresponds to a half-integral number of wavelengths, cancellation will occur, and there will be no intensity registered at the detector.

Example. A set of planes in a crystal is separated by 250 pm ($2.50 * 10^{-10}$ m). What is the maximum x-ray wavelength that could be used to measure this distance?

Solution. The largest scattering angle possible is 90°. Since sin 90° = 1, n = 2d = $5.00 * 10^{-10}$ m. q is largest for n = 1, giving = $5.00 * 10^{-10}$ m. A wavelength longer than this cannot fit even once into the distance, 2d, so cannot be used to measure the distance. X-rays with less than $5.00 * 10^{-10}$ m must be used.

LATTICE IMPERFECTIONS

Perfect crystals, in which every atom or molecule is in place, are an idealization, like the ideal gas. Real crystals contain fairly large numbers of imperfections, called defects. Three major types of defect are called point defects, line defects, and plane defects. Common point defects in simple lattices (in which all lattice points are identical) are vacancies (in which an atom is missing); atoms out of place, occupying what is normally a hole in the lattice; substitutional impurities, in which foreign atoms occupy occasional sites; and interstitial impurities, in which foreign atoms occupy holes in the lattice.

Ionic (compound) lattices exhibit Schottky defects, in which a cation or anion is missing; and Frenkel defects, in which an ion leaves its site and enters a lattice hole. Line defects result from missing rows of atoms. Particularly common are edge defects, in which an extra partial plane of atoms is inserted near an edge of the crystal.

Example. Bending a paper clip. Take an ordinary paper clip and try to straighten it out. Are you able to get it perfectly straight in the regions where it was bent? Why not? Now try to rebend the paper clip to its original shape. Are you able to do it? Why not?

Solution. When the clip was manufactured, it was annealed following bending. Annealing is the process of heating the clip to soften it enough that atoms can diffuse into the gaps left in the bending process. In other words, annealing produces edge defects. When you attempt to straighten the clip, the edge defects prevent the alignment of atomic planes, so complete straightening is impossible. However, straightening the clip introduces new edge defects on what was originally the inside curve of the bend, so that when you attempt to rebend the clip, the new defects interfere.

Plane defects in crystals occur at the interface between two differently-oriented crystalline regions. As might be expected, such defects greatly diminish the resistance of the crystal to fracture. Crystal defects are important for a number of reasons. A problem with defects, especially of the line and plane type, is that they diminish the structural strength of the solid. Annealing is used to offset this. Rusting of iron often occurs at defect sites on the surface, which may result when the metal is worked or penetrated by a rivet or bolt. However, defects can be advantageous as well.

Defects are sites of high energy (*i.e.*, relative instability). Reactions that take place at the surfaces of crystals often take place at these sites. This is involved in heterogeneous catalysis by solid materials, which is used on a huge industrial scale in our economy. Electronic components for modern computers require the reliable production of silicon that has a purity of

99.999999%. Silicon crystals of such purity contain very few defects, which would diminish their functional effectiveness. Silicon of this purity is produced by a process called zone refining. In this method, a bar of silicon of high purity is passed very slowly past a very narrow heating element, which melts a narrow region of the silicon bar. Impurities in the silicon dissolve in the melted part, and follow along with the melted region as the bar moves past the heating element. Eventually the fused region, containing all of the impurities, reaches the end of the bar, where it is solidified and cut off. The remaining silicon has the purity required in the manufacture of electronic components.

PROPERTIES OF LIQUIDS

We take up the liquid phase last because it is the most complex and least understood of the phases of matter.

Liquids have properties intermediate between those of the solid and gas phases:

- A sample of liquid has a fixed volume (like a solid) and a molar volume only slightly greater than that of the solid. However,
- A liquid conforms to the shape of its container, like a gas, and
- A liquid flows readily, like a gas.
- Liquids are relatively incompressible. The relative compressibilities of the three phases are in the order g >> l > s. The incompressibility of liquids finds practical use in hydraulic machinery, such as automobile brakes.
- Diffusion is slower in liquids than in gases, but faster than in solids. The relative rates of diffusion in the three phases are in the order g >> l > s.

These properties are readily observed visually.

Several other properties of liquids are more subtle:

- Liquids give x-ray diffraction patterns that are diffuse, rather than sharp like those of solids. The pattern implies some order in the molecular arrangement, but less than in the solid.

- Liquids exhibit surface tension.
- Liquids exhibit viscosity.
- Liquids in closed containers evaporate to an extent. The resulting vapour exerts a pressure called the vapour pressure, P_{vap}, of the liquid. The magnitude of P_{vap} for a particular liquid depends only on temperature, not on the container size or the amount of liquid present. Liquid and vapour exist in equilibrium.

We begin with a molecular model for the liquid phase, based on its observable properties.

MODEL OF THE LIQUID PHASE

The small molar volume and incompressibility of liquids suggest that the molecules are close together. However, flow and the lack of a fixed shape indicate molecular mobility, despite the closeness. As for the gas and the solid phases, the kinetic energies of the particles of the liquid are described by the Maxwell-Boltzmann distribution, and the average KE per particle is 3/2 kT. Each molecule is fairly closely surrounded by other molecules, but there are definite gaps (holes) in the structure. Molecules use these gaps to slip and slide easily past one another, which manifests macroscopically in the ability to flow.

There are regions in the liquid that are quite ordered, similar to the solid, but the regions are constantly shifting position as molecules move to close gaps and open new ones. The gaps prevent the order from extending over long distances. Thus liquids have short-range order, in contrast to the long-range order of crystalline solids. The gaps allow diffusion of a molecule through the liquid, but frequent collisions with neighboring molecules makes diffusion slow.

Contrast this with the rapid diffusion in the gas phase, where a molecule travels a long distance between collisions; and the extremely slow diffusion of the solid phase, where a molecule is locked into its lattice location. Diffusion rates in solid, liquid, and gas are best understood in terms of the mean free path, the average distance travelled by a molecule between collisions. The mean free path in the solid is virtually zero. In the liquid, it is a fraction of the molecular diameter. But in the gas, the mean free path may be 10-100 molecular diameters, depending on gas pressure.

INTERESTING PHYSICAL PROPERTIES. DIFFUSE X-RAY DIFFRACTION PATTERNS

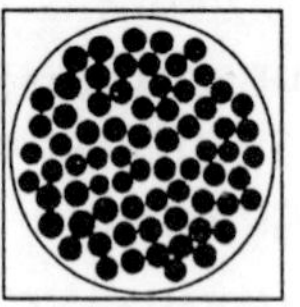

Fig. 9.22

Figure 9.22 indicates regions of short-range order in the liquid. Within these regions, molecules are almost close-packed. However, the gaps allow fluctuation in the distance of separation of the molecules. The nearly close-packed regions of the liquid diffract x-rays according to the Bragg equation, n = 2dsintheta. Fluctuations in d, a consequence of the gaps, cause fluctuations in theta, giving x-ray patterns that are diffuse, or smeared out.

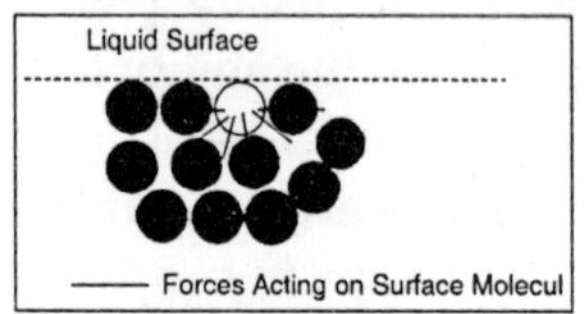

Fig. 9.23

SURFACE TENSION

This is a consequence of intermolecular forces. Consider a molecule at the surface of a liquid. In contrast to a molecule in the bulk, which is attracted equally from all directions, the surface molecule suffers an imbalance of forces because there are no molecules above it. The net force on it is directed towards the interior of the liquid. This pulling in of the liquid surface is called surface tension, symbolized. The surface molecule has higher potential energy than a bulk molecule because of the lesser force on it. To minimize energy, the liquid adopts a shape that minimizes the number of surface molecules.

The shape with the smallest surface area–to–volume ratio is the sphere. Thus surface tension explains the spherical shape of raindrops and water droplets on a freshly waxed car: the spherical shape minimizes potential energy. It is a safe prediction that manufacture of ball bearings will someday be carried out in space, where gravitational forces will not distort the sphericity that surface tension imposes on a drop of molten metal. As temperature increases, molecular kinetic energy can overcome the attraction of intermolecular forces, and surface tension decreases according to equation 7–8–1.

$$(7\text{–}8\text{–}1): \zeta = a - bT$$

The constants a and b are characteristic of a particular liquid. For ethanol, $a = 24.05 * 10^{-3}$ J/m^2 and $b = 0.0832 * 10{-}3$ J/m^2-°C. A drop of water on a clean glass surface spreads and flattens, forming a film on the surface. This phenomenon of spreading and flattening is called wetting. Wetting occurs because forces between the surface molecules and molecules of the liquid exceed the intermolecular forces within the liquid. The liquid spreads to maximize the area of contact with the surface. Forces between unlike molecules are called adhesive forces; those between like molecules are called cohesive.

Wetting occurs because adhesive forces exceed cohesive forces. If the reverse is true, a liquid will not wet a surface, but instead will sit on the surface in spherical droplets to minimize the area of contact with the surface. One visible consequence of wetting is the formation of a meniscus at the top of a column of liquid in a tube, such as a buret or pipet. Figure 9.24 shows the menisci formed by water and mercury. The adhesive forces between the glass surface and water molecules are stronger than the cohesive forces between water molecules. Consequently, water achieves a maximum surface area of contact with the glass by bending up at the edges and sagging in the center. The reverse is true for mercury, which minimizes the surface area of contact with the glass and forms a spherical surface shape to maximize cohesive forces.

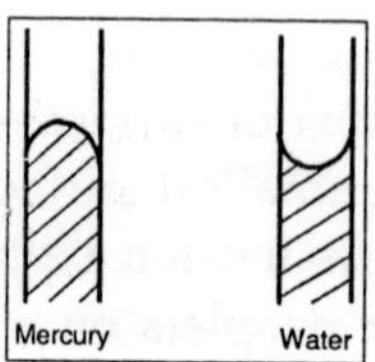

Fig. 9.24

The phenomenon of capillary action results from competition between surface tension and wetting. When one end of a capillary tube is submerged in a liquid, a difference in height develops between the levels of liquid in the capillary and in the bulk liquid. A liquid that wets glass rises in the tube; a liquid that does not wet glass falls. Figure shows this behaviour, called capillary action, for water and mercury.As water wets the inside wall of the capillary, the surface area of the water is increased. However, surface tension tends to minimize the number of molecules at the surface, and causes the water to move up the tube, against the force of gravity, to offset the surface area increase resulting from wetting. The result is a capillary rise, as in Figure. For mercury, adhesive forces are weak, so the level in the capillary falls to minimize contact with the glass surface, and the surface of mercury in the capillary is pulled into a sphere by the strong cohesive forces.

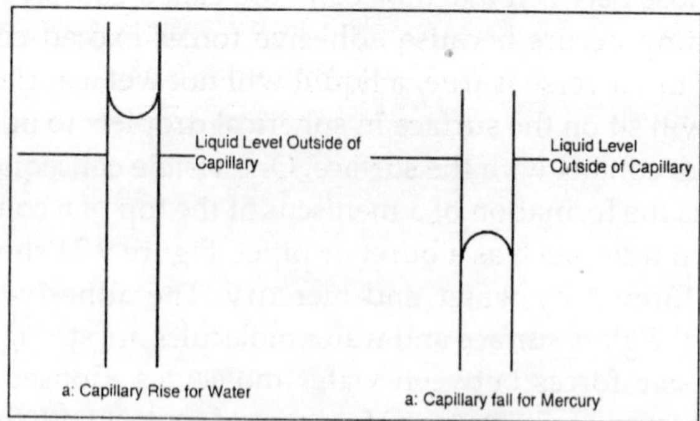

Fig. 9.25 Capillary Action

The result is capillary fall, as in Figure 9.25. Capillary action is important in the operation of blotters and sponges and is the mechanism by which water penetrates soils.

VISCOSITY

The viscosity of a liquid is a measure of its resistance to flow. A mobile liquid is one that flows readily (*e.g.*, water). A liquid that flows with difficulty (*e.g.*, molasses) is said to be viscous. Liquids composed of small, roughly spherical molecules are mobile, because the molecules can roll over one another smoothly, like ball bearings. Viscous liquids are composed of long, string–like molecules that easily become tangled during flow. For example, the molecules of high viscosity motor oils consist of long chains of carbon atoms that intertwine and impede flow of the liquid. An extreme example of molecular tangling occurs in silly putty, a polymer that appears to be a solid but is in reality a liquid in which molecular tangling is so severe that hours are required for the putty to spontaneously change shape to that of the container. Like surface tension, viscosity decreases as temperature increases due to more rapid thermal motion of the molecules.

VAPOUR PRESSURE

Vapour pressure is the most fascinating physical property of liquids. It provides remarkable insight into the laws governing the behaviour of physical systems.We begin with a purely experimental approach to the phenomenon.

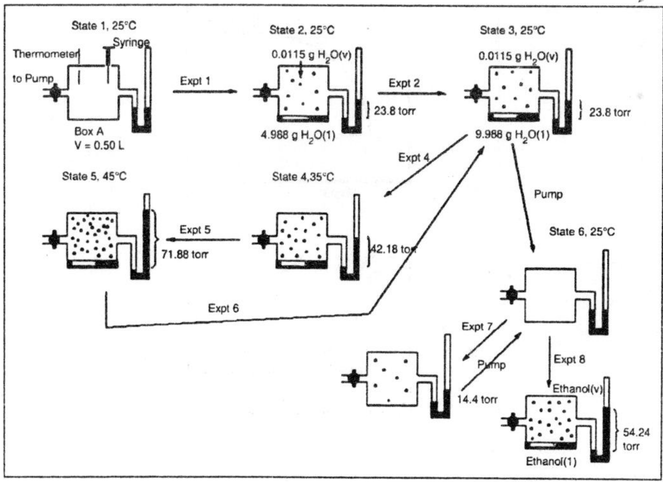

Fig. 9.26

A thermometer is inserted in the box and there is a septum–capped port (opening) that can be penetrated with a syringe needle. The stopcock to the vacuum pump is opened and the air is pumped out until the two arms of the manometer are at the same level. The stopcock is then closed. This is the initial situation, state 1, shown at the top left of the figure 9.26. Several experiments will now be performed in sequence.

Expt 1. The box is thermostatted at 25°C, and 5.000 g (approximately 5 mL) of water is injected through the septum by syringe. We observe that a pressure develops within the box. Over a brief time (perhaps 1 minute) the pressure rises to 23.8 torr and remains there. The change of pressure with time is as shown inFigure. Note that the rate of change of the pressure (the slope of the curve) is largest just after injection of the liquid, and gradually decreases as the pressure smoothly approaches the final value.

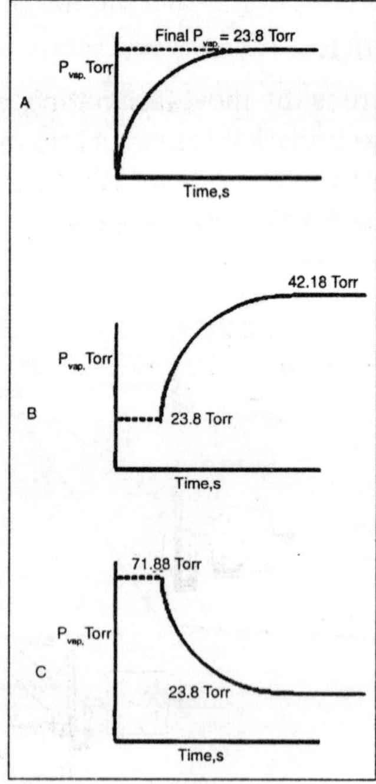

Fig. 9.27 Vapour Pressure Dynamics

There is still liquid at the bottom of the box. This is the situation at state 2 in Figure 9.27. We conclude that some water has evaporated to form water vapour, which is responsible for the observed pressure. Since the pressure is caused by the vapour formed by evaporation of liquid, it is called the vapour pressure, P_{vap}. Assuming that water vapour behaves as an ideal gas, we can use the ideal gas law to calculate the moles of water in the vapour phase. The result is 6.40×10^{-4} moles, or 0.0115 g, of water in the vapour phase.

Expt 2. With T = 25°C and the vapour pressure steady at 23.8 torr, we inject 5.000 g more liquid water to the box. This time we observe no change in the pressure registered at the manometer. This is the situation at state 3 of the diagram. We conclude that the value of P_{vap} does not depend on how much liquid water is in the box.

Expt 3. This experiment will be done using the box in Figure 9.28. It is identical in all respects to the first box except that its volume is 1.00 L. In state 1, Figure 9.28, the 1.00–L box has been evacuated by pumping until the Hg level in both arms of the manometer is the same. With the box at 25°C, we inject 5.000 g H_2O.

The moles of water in the vapour phase must be twice as great as for the original box, because the volume is twice as great. But the pressure is the same. We conclude that P_{vap} is independent of the container size.

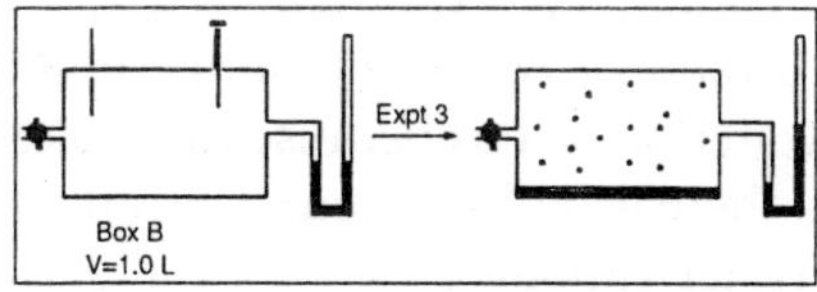

Fig. 9.28 Vapour Pressure Experiments

Experiments 4-6 tell us that P_{vap} increases with temperature. Moreover, the increase is non-linear: the increase between 35 and 45° (29.7 torr) is much greater than that between 25 and 35° (18.38 torr). We will do one last experiment using water. Remove all liquid water from the 0.500–L box, and pump out all water vapour. The box is empty with equal levels of Hg in the manometer arms. This is state 6. We inject 0.00700g H_2O into the box. This time, the pressure increases to a final value of 14.4 torr and all of the liquid disappears. We conclude that after injection, evaporation occurs as the water attempts to achieve a pressure of 23.8 torr, the characteristic P_{vap} at 25°C.

However, before this pressure can be reached, all of the water has evaporated. The amount of liquid injected was insufficient to achieve the characteristic P_{vap}. The minimum quantity of water needed to achieve P = 23.8 torr was found in Expt 1 to be 0.0115 g. Any amount less than this will evaporate entirely, giving a final pressure equal to the value calculated using the injected number of moles in the ideal gas law.

Empty the 0.500–L box and inject 5.000 g of a different liquid, say ethanol (ethyl alcohol, C_2H_6O) at 25°C. We observe the same pattern of behaviour as with water, except that the final vapour pressure is 54.24 torr. We conclude that all liquids have a tendency to evaporate to some extent in a closed container, and establish a vapour pressure that depends only on T and the identity of the liquid. The vapour pressure phenomenon is a universal property of liquids.

Before summarizing the results of our thought experiments, we deal with a related matter. Suppose we put 10 mL of water in a beaker and leave it open to the atmosphere. Over a few days, the water evaporates completely. (Acetone and ethyl alcohol would completely evaporate in a few hours; diethyl ether in a few minutes.) Why? This situation differs from that involving the water in the box because the beaker system is not closed—that is, there is no fixed boundary on top of the beaker to contain the water vapour.

Water in the beaker evaporates in an attempt to establish the characteristic Pvap (23.8 torr at 25°C). Water vapour develops over the surface of liquid in the beaker, but begins to diffuse away because there is no confining boundary on top. This diffusion, aided by air currents in the room, moves vapour molecules away from the space over the beaker. This reduces the local water vapour pressure, and more water evaporates to make up for the loss. This process continues until all liquid is gone. We will see in that this is a situation in which equilibrium cannot be attained between liquid and vapour.

The results of our experiments with vapour pressure can be summarized as follows:

- A liquid in a closed container evaporates to some extent. The resulting vapour exerts a characteristic vapour pressure, Pvap.
- The value of Pvap is independent of the container size and the amount of liquid, provided that the amount exceeds a critical minimum.
- Pvap increases steeply with T.
- Pvap is a function only of the chemical identity of the liquid and the temperature.
- Liquids have a natural tendency to evaporate to some extent to form vapour.

The thought experiments, which can be verified readily in the laboratory, provide much descriptive information about the vapour pressure phenomenon. But the questions of why liquids tend to evaporate, and why they establish characteristic vapour pressures, are unanswered by the experiments. Answers to these questions are forthcoming in, where we attempt a molecular (microscopic) interpretation of vapour pressure and introduce the concept of phase equilibrium.

THE DISSOLVING POWER OF LIQUIDS

Liquids are invaluable to chemists as solvents. They are able to accommodate molecules of solute readily within the gaps in their structures to produce solutions that are themselves liquids. Such solutions are readily poured or otherwise transferred from one container to another; are conveniently measured out quantitatively using volumetric apparatus; and provide a uniform medium in which a chemical reaction can occur. The unique ability of liquids to act as solvents is due to their dissolving power the molecular motion of the liquid dislodges ions or molecules from the surface of crystals of a solid, then accommodates the separated ions or molecules in gaps.

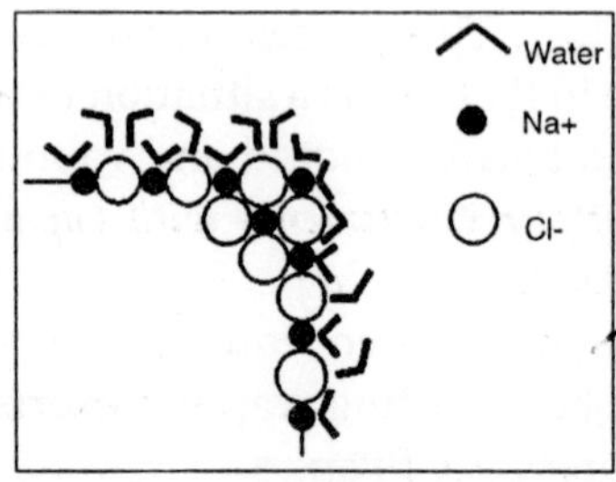

Fig. 9.29

Figure 9.29 shows a microscopic view of two adjoining faces of a crystal of NaCl. A Na^+ ion on a face of the crystal differs from those in the interior because it experiences forces of attraction from only 5 nearest-neighbour Cl^- ions, rather than 6. Facial Cl^- ions are in a similar situation. Such ions are therefore in shallower potential wells than ions in the interior of the crystal and should be more easily dislodged. Edge ions, such as the Na^+ ion labelled "a" in the figure, are attracted by only 4 nearest neighbours, and should be even more readily dislodged than face ions. Water molecules from the solvent continually bump against the faces and edges of the solid crystal. As they approach the crystal, their dipoles become oriented due to interactions with the face ion that they approach. Water molecules approaching cations have the O atom oriented towards the crystal face; those approaching anions put a H atom forward.

When a rapidly moving water molecule strikes a face or edge ion, sufficient kinetic energy can be transferred to dislodge the ion, which moves away from the crystal into a gap in the solvent structure. It becomes immediately surrounded by a cage of water molecules, with oxygen atoms nearest the Na^+ ion to maximize ion-dipole forces. The solvent cage insulates it from the crystal and makes return to the crystal face unlikely. Since edge ions are more readily dislodged than facial ions, and because they can be approached by solvent from several directions, dissolution occurs more rapidly at crystal edges than faces. The initially sharp edges of a dissolving crystal become rounded as dissolution proceeds.

PHASE TRANSFORMATIONS

A phase transformation is a conversion of a pure substance from one phase to another.

There are 6 types, each involving two of the three phases:

1. *Melting (fusion):* Conversion of solid to liquid.
2. *Freezing:* Conversion of liquid to solid; the reverse of melting.
3. *Vaporization:* Conversion of liquid to vapour (gas). Its reverse is
4. *Condensation:* Conversion of vapour to liquid.
5. *Sublimation:* Conversion of solid to vapour. Its reverse is
6. *Deposition:* Conversion of vapour to solid.

The phase changes are illustrated in Figure. Since the solid and liquid phases have small molar volumes due to close packing of molecules, they are referred to as condensed phases.

The two variables that determine the phase in which a substance exists are the temperature and the pressure to which the substance is subjected. Phase transformations may therefore be accomplished in two limiting ways: by changing the temperature of the substance while keeping it under constant pressure; or by changing the pressure on the substance while keeping it at constant temperature. The first approach is the more familiar of the two. We discuss that now and postpone a discussion of pressure-induced phase changes.

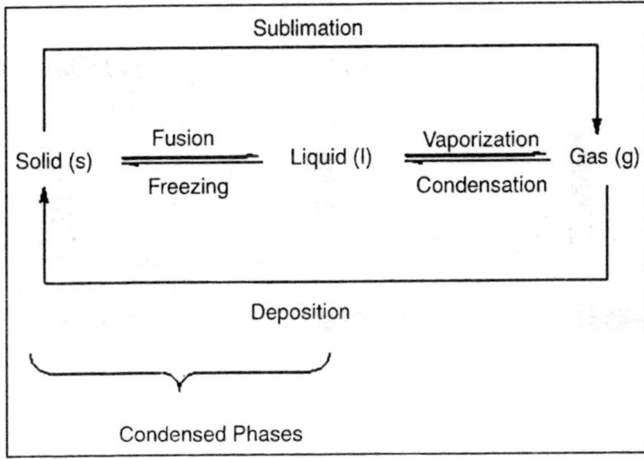

Fig. 9.30

We make three assertions about temperature-induced phase changes:

1. At a constant pressure, a phase transformation occurs at a constant temperature that is characteristic of the substance.
2. To convert a given amount of substance from solid to liquid to gas requires addition of energy. This is usually accomplished by adding heat. The energies of the phases are therefore in the following order: $E_s < E_l < E_g$
3. Conversion from solid to liquid to gas decreases the order (degree of organization) of the substance.

To clarify these assertions, we take a thought experimental approach to the phase changes of a substance, much as we did for vapour pressure. The results of these thought experiments can be readily verified in the laboratory. We start with a quantity of solid water (ice) in a syringe, as shown in Figure 9.30, state 1. The initial temperature of the ice and the inside of the syringe is –10°C. We plan to slowly add energy to the ice by controlled heating, which can be done by submerging the syringe in a large antifreeze bath with a hot plate under it.

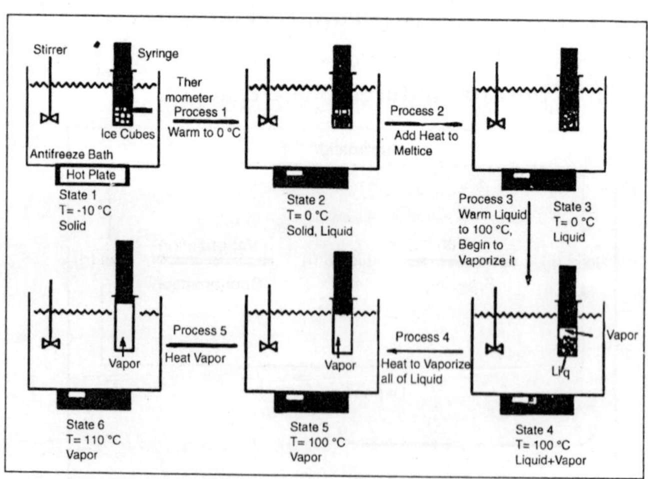

Fig. 9.31

As we slowly add heat, we monitor the temperature of the

water in the syringe with a thermometer inserted through an air-tight seal in the syringe. We begin at state 1. Process 1. Turn on the hot plate and the bath stirrer. As the ice warms slowly from –10 to 0°C, nothing seems to happen. Temperature changes slowly and steadily. When the ice reaches 0°C, it begins to melt, and we see some liquid water form in the syringe. This is state 2 of Figure 9.31.

Process 2. We continue to gently heat the bath. The ice gradually melts, but the temperature of the ice-water mixture remains constant at 0°C, despite the addition of heat. The temperature of the bath rises above 0°C, but as long as both ice and liquid water are in the syringe, T in the syringe remains at 0°. State 3 is reached when the last ice crystal melts. All water in the syringe is liquid, at T = 0°C.

Process 3. Once the ice has melted, the temperature of the water in the syringe begins to rise. As we slowly warm the bath, the water temperature slowly rises. Nothing happens in the syringe other than a slight increase in the volume of the liquid until T = 100°C. Then water begins to vaporize. This is the situation at state 4.

Process 4. We continue to slowly add heat to the bath and observe that temperature in the syringe remains constant at 100°C as long as both liquid and vapour are present, even though the bath temperature rises above 100°C. We also observe that as vapour forms, it pushes the plunger out. After all liquid has vaporized, at state 5, the temperature of the water in the syringe once more starts to rise.

Process 5. Continue heating the bath to T = 110°C. The temperature of the vapour in the syringe rises from 100° at state 5 to 110° at state 6, while its volume increases according to the ideal gas law. The pressure of the water vapour is maintained constant at 1 atm by movement of the syringe plunger.

First, phase changes occur at constant T. This is shown by processes 2 and 4. The fixed, characteristic temperature at which a substance melts is called the melting point (s to l) or freezing point (l to s), and is symbolized T_f. The normal melting point is the temperature of melting under a pressure of 1.00 atm. For water, this is 0°C. The characteristic temperature at which a substance boils under 1.0 atm pressure is called the normal boiling point. For water, this is 100°C. Second, process 2 shows that an input of energy is required to convert a solid to liquid. The same is true for conversion of liquid to gas (process 4). For each of these processes, the change in energy, ΔE ($E_{final}-E_{initial}$) is positive. Processes 2 and 4 may be represented as follows:

$$\text{(7–10–1)}: H_2O(s), 0°C \rightarrow H_2O(l), 0°C \; [\Delta E > 0]$$

$$\text{(7–10–2)}: H_2O(l), 100°C \rightarrow H_2O(g), 100°C \; [\Delta E > 0]$$

The process in equation 7–10–1 occurs at constant T. Therefore the change in kinetic energy of the molecules is 0 (recall that an average molecule has KE of 3kT/2 at temperature T, regardless of phase). ΔE is therefore a change in potential energy, ΔPE. Thus the PE of liquid is higher than that of solid water at 0°C. Similarly, the PE of vapour is higher than that of liquid water at 100°C. Combining these conclusions gives equation 7-10-3:

$$\text{(7–10–3)}: PE_g > PE_l > PE_s \text{ at any T}$$

$\Delta PE > 0$ for sublimation, fusion, and vaporization; DPE < 0 for the reverse processes. Third, the amount of disorder at the molecular level increases in our sequence of processes (for reasons that will become clear later, we phrase this in terms of disorder rather than order). With addition of thermal energy (heat), the molecules convert from the long–range organized arrangement of the solid to the completely chaotic and random molecular distribution of the gas. Even warming the gas from 100 to 110°C decreases the order, because the volume available for the molecules to bounce around in increases. This conclusion is summarized in equation 7–10–4.

PHASES AND THE POTENTIAL WELL, REVISITED

Discussed the origin of the potential well in terms of electromagnetic forces within or between the fundamental units of substances. We concluded that forces are strongest and potential energy is minimized when fundamental particles form an orderly close-packed arrangement, and we recognized this as the situation in the solid phase. Heating the solid phase increases the kinetic energies of the particles of the solid phase until they are able to partially overcome the close-packed forces and separate somewhat.

At this point, solid melts to liquid. Separation has increased the potential energy of the particles. Further heating eventually gives the particles sufficient kinetic energy that they may escape the forces in the liquid into the space above the liquid; *i.e.*, vaporization occurs. In the vapour phase, molecules are so widely spaced that forces between them are negligible and potential energy has increased to zero. We thus see that phase conversion from solid to gas can be viewed in terms of particles climbing out of the potential well. Similarly, in the reverse conversion, particles fall into the potential well. The decrease in potential energy of the molecules is released as heat.

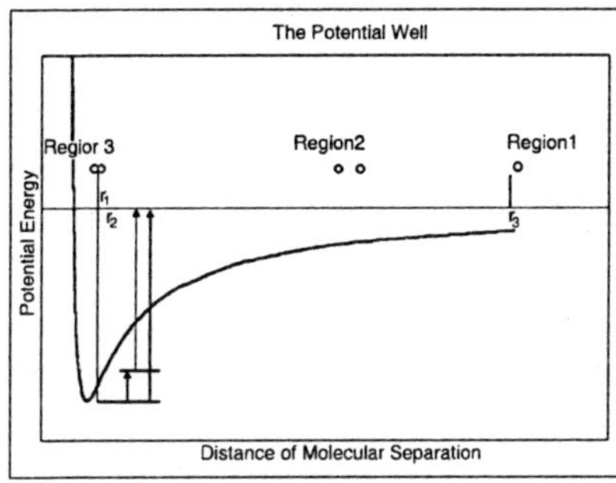

Fig. 9.32

APPLICATIONS

In the 2–dimensional lattice, outline a primitive unit cell and a non-primitive unit cell. Assuming that there is an atom at each lattice point, calculate the number of atoms contained within each of your unit cells.Draw a simple cubic unit cell; a body-centered cubic unit cell; a face-centered cubic unit cell. Show how to determine the number of atoms per unit cell in each lattice. Give the coordinates of the tetrahedral holes in a fcc unit cell.

Give the coordinates of the octahedral holes in a fcc unit cell. Determine the number of atoms per hexagonal unit cell. Show that in a face-centered cubic lattice, 74% of the available space is occupied by atoms. What fraction of the available volume is occupied by atoms in a simple cubic unit cell? In a body-centered cubic unit cell? Calcium crystallizes in a face-centered cubic lattice. The length of the edge of the unit cell is 0.556 nm (1 nm = 10^{-9} m). What is the radius of a calcium atom? Sodium oxide, Na_2O, has a lattice consisting of fcc O^{2-} ions with Na^+ ions in the tetrahedral holes. Show that there are 4 formula units of the compound per unit cell. Show that the coordination number of Na^+ is 4, whereas that of O^{2-} is 8.

How many carbon atoms are there per unit cell of the diamond structure? What is the coordination number of each C atom A set of planes in a crystal is separated by 645 pm. What is the longest x–ray wavelength that can be used to measure this distance? Iron(II) oxide is an example of a non-stoichiometric compound. This means that the ratio of iron to oxygen atoms is non-integral. This situation can occur whenever the cation can exhibit more than one stable oxidation state.

In this case, iron exhibits oxidations states +2 and +3. A sample of iron(II) oxide is prepared with the composition Fe0.85O. What fraction of the iron atoms are Fe(III)? A crystal of very pure silicon suitable for semiconductor devices contains only 0.00001% defects. Silicon has the same crystal structure as diamond. How many unit cells contain defects in a 1-mole crystal of silicon? When a perfectly cubic crystal of sodium chloride, NaCl, is placed in water, it begins to dissolve. As it does so, the corners and edges of the cube become rounded, but the faces remain flat. Explain these observations in atomic/molecular terms. A particular set of planes in a crystalline solid has a spacing distance of $4.38 * 10^{-10}$ m (0.438 nm). X–rays of wavelength 0.1886 nm are diffracted from this set of planes. At what angles will reflections from these planes be observed?

Explain each fact:

- Motor oil is viscous at 25°C, but flows readily in an operating engine.

- Water bugs are denser than water, but can scoot around on its surface.
- A glass can be filled "fuller than full" if water is added carefully.
- Tap water beads up on a dirty glass but forms a film on a clean glass.
- Wet grains of sand clump together.

Spontaneous mixing of two liquids is more rapid at high than at low T. Explain this observation. 7–18. Relative humidity indicates the water content of the atmosphere. 50% relative humidity at a certain T means that the partial pressure of $H_2O(v)$ is 50% of the vapour pressure of water at that T. In which case is the partial pressure of water vapour larger: 50% relative humidity at 25°C, or 40% relative humidity at 30°C? $P_{vap}(H_2O)$ = 23.8 torr at 25°C, 31.8 torr at 30°C. 2.0 g liquid H_2O is injected into an evacuated box with V = 0.521 L. The pressure in the box rises to a final value of 23.8 torr at 25°C. Calculat

- Moles of water in the vapour phase.
- Moles of water in the liquid phase.

A 0.750–L glass bulb contains a mixture of He(g) and $H_2O(g)$ at 22°C and a pressure of 52 torr. A side arm on the bulb is cooled to –195°C and the water vapour is frozen out completely as $H_2O(s)$. $H_2O(s)$ has a negligible vapour pressure at –195°C. He remains as a gas. The pressure in the bulb is then found to be 18.5 torr. How many moles of He are in the bulb and what was the partial pressure of water at 22°C? A gas mixture in a 0.500–L bulb contains $N_2(g)$, $H_2O(g)$, and $CO_2(g)$ at a total pressure of 632 torr at 25°C. $CO_2(s)$ has a high vapour pressure (1.013 bar) at dry ice temperature (–78°C) but the vapour pressure of water is negligible at –78°C.

Neither $CO_2(s)$ nor $H_2O(s)$ has a significant vapour pressure at the temperature of liquid nitrogen (–195°C), but $N_2(l)$ has a vapour pressure of 1.013 bar (atmospheric pressure) at -195°C. When the bulb sidearm is cooled to –78°C, the pressure falls to 610 torr. When the sidearm is cooled to –195°C, the pressure falls to 18.0 torr. How many moles of each gas are in the bulb? P_{vap} of $H_2O(l)$ at 0°C is 4.54 torr. 5.00×10^{-3} g of $H_2O(l)$ is frozen at a very low temperature in the bottom of the sidearm tube of a gas sample bulb of volume 1.5 L. All of the air is pumped out and the stopcock is closed. The water at the bottom of the tube is melted by heating it to 0°C. The vessel is kept at 25°C. What phases of H_2O will be present at the end of this process, and how much of each will there be? $MM(H_2O)$ = 18.015 g/mole. 0.0082 g of ethyl alcohol, C_2H_5OH, is injected into an evacuated 0.493–L bulb at 31°C. P_{vap} of ethyl alcohol is 74.86 torr at 31°C. What will be the final pressure in the bulb? What mass of liquid ethanol will be present? A liquid is in equilibrium with its vapour in a cylinder/piston apparatus.

Show graphically how the rates of evaporation and condensation vary with time:

- From time t_0 to time t_1, the system is at equilibrium.
- At time t_1, T is instantly increased to some higher value. The system returns to equilibrium.
- At time t_2, the piston is instantly moved down to point B. The system returns to equilibrium in the smaller volume.
- At time t_3, T is instantly decreased. The system returns to equilibrium.

5.00 g of $H_2O(l)$ is injected into a 1.50 L box at 35°C. $P_{vap}(H_2O)$ at 35°C = 42.18 torr. Calculate the mass of H_2O in the g and l phases after P_{vap} has been established.

A liquid is in equilibrium with its vapour in a cylinder/piston apparatus.

- From time t_0 to time t_1, the system is at equilibrium.
- At time t_1, T is instantly decreased. The system returns to equilibrium.
- At time t_2, the piston is instantly moved up to point A. The system returns to equilibrium in the larger volume.
- At time t_3, additional vapour is instantly injected into the cylinder. The system returns to equilibrium.
- At time t_4, some vapour is instantly withdrawn from the cylinder. The system returns to equilibrium.

The compound, Na_3AlF_6, is called cryolite. It is important in the production of aluminum.

- Cubic close packed F^- ions with Na^+ ions in the octahedral holes and Al^{3+} in the tetrahedral holes.
- Cubic close packed AlF_6^{3-} ions with Na^+ ions in all of the octahedral and tetrahedral holes.
- Cubic close packed Na^+ and Al^{3+} ions with F^- ions in the octahedral holes.
- Cubic close packed Na^+ ions with AlF_6^{3-} ions in the tetrahedral holes.

Chapter 10

Principles of Chemical Dynamics

INTRODUCTION

- Rates of chemical reactions depend upon the structures and concentrations of the reactants and products; the presence or absence of a catalyst; and the temperature.
- Reaction rates increase exponentially with temperature, as molecular kinetic energy increases according to the Maxwell-Boltzmann Distribution.
- Reaction rates increase in the presence of appropriate catalysts.

SOME BASIC IDEAS

Until now, our major concern has been whether or not a chemical reaction of interest may or may not occur under specified conditions of temperature, pressure, and concentration. The answer to this question is in the realm of thermodynamics. If thermodynamics tells us that a reaction may occur, then we must be concerned with how rapidly and by what pathway it will occur, and how we may influence this. The speed (rate) at which a reaction takes place is a matter of *chemical dynamics*, or alternately, *chemical kinetics*.

Chemical dynamics (kinetics) is the study of the rates (speeds) and mechanisms (pathways) of chemical reactions. We have briefly encountered the rate concept before in, when we discussed the manner in which vapour pressure changes with time; and in, when we presented dynamic equilibrium as a situation of balanced, opposing rates. Reaction rates will be our chief concern in this chapter. Although chemical kinetics is a well-developed field experimentally, we are still a long way from the theoretical grasp of rate that we have of equilibrium. Thus our predictive ability in the realm of kinetics is very limited.

Thermodynamics may predict that a reaction should proceed essentially completely. Whether it does so depends on the rate with which it occurs. For example, consider the reaction below at 298 K:

$$C(graph) + O_2(g) \rightarrow CO_2(g) \; [\Delta G^o_R = -394.4 \text{ kJ}]$$

Thermodynamics predicts that the reaction should go to completion. But carbon (coal) deposits exist in the atmosphere without noticeably reacting; and the "lead" in our pencils does not burst spontaneously into flame. The reaction is favoured, but is very slow at 298 K. We say that C(graph) is thermodynamically unstable in the presence of $O_2(g)$ at 298 K, but is kinetically inert. As a second case, consider the acid-base neutralization reaction:

$$H_3O^+(aq) + OH–(aq) \rightarrow 2H_2O(l) \; [\Delta G^o_R = -58.28 \text{ kJ}]$$

Thermodynamics predicts that this reaction, too, should go to completion. In fact, it occurs almost instantaneously; equilibrium is attained as rapidly as the reactants are mixed. Thus we say that H_3O^+ is both thermodynamically unstable and kinetically labile in the presence of OH^-. Two sets of descriptive words are necessary to distinguish thermo-dynamics and kinetics:

Thermodynamics: stable $\Delta G > 0$; unstable $\Delta G < 0$

Kinetics: inert (reaction is slow); labile (reaction is fast)

As we have seen above, there is no necessary correlation between "stable" and "inert", or "unstable" and "labile". This lack of correlation reflects the fact that thermodynamics deals with values of state functions in various equilibrium states of a system; kinetics deals with the rate at which a path between states is traversed. Reaction rates are extremely important in many ways in our lives. Sustained muscle action depends on the continuous supply of energy produced by glucose combustion in the mitochondria of cells. Energy production must be fairly rapid, otherwise sustained action is not possible. The extremely rapid (explosive) burning of gasoline in the internal combustion engine provides the motive power for automobiles.

Food cooks (or, as the chemist would say, oxidizes) extremely slowly at room temperature, but in a relatively short time at more elevated temperature. Our first concern must be to discuss the manner in which reaction rates are expressed, and to introduce some of the terminology of chemical dynamics.

What is a *rate*? In general, a rate is a number expressing how much a quantity changes in a convenient interval of time. Thus an interest rate is the amount by which a sum of money increases in 1 year; travel rate in an automobile or airplane is the distance covered in 1 hour; heart rate is the number of heartbeats per minute; thus a rate is the change in a quantity per unit time:

$$\text{Rate} = \Delta\text{quantity}/\Delta t$$

Note that the units of rate must be "quantity/time, or "quantity–time^{-1}". Thus interest rate has units of dollars/year; travel rate, miles/hour; and heart rate; beats/minute.

How are we to express the rate of a chemical reaction? To develop the answer to this question, we will begin with the simple chemical process in equation 15–1–2. The equation describes quantitatively the conversion of ozone to oxygen that occurs in the earth's upper atmosphere.

$$2O_3(g) \rightarrow 3O_2(g)$$

When this (or any other) reaction takes place, it is the amounts of reactants and products that change. As we know, amount can be expressed in either mass, moles, or molecules. The coefficients in the chemical equation relate both molecules and moles of reactants and products but do not directly relate masses. So we will agree that our measure of amount should be either molecules or moles. Further, since work in the laboratory is carried out on the macroscopic scale, moles is the more suitable unit of amount. So we might describe the rate of process 15–1–2 either in terms of the number of moles of O_3 used or of O_2 produced per unit time. However, there is a major difficulty with this approach. The number of moles of ozone used will depend upon the volume of atmosphere that we choose to examine!

We cannot specify a rate without also specifying the volume examined. Rather than use moles/time, then, we choose to divide by the appropriate volume to give moles/liter-time, or concentration/time. Rates expressed in these units are independent of the actual amount of reaction carried out. At this point then we can express the rate of reaction 15–1–2 in either of two ways:

(15–1–3a): Rate = Moles O_3 used per liter per time
= –Δ(conc O_3)/Dtime

(15–1–3b): Rate = Moles O_2 produced per liter per time
= Δ(conc O_2)/Dtime

The units of the rate are then moles per liter per time, where the time unit should be the one most convenient for the reaction being measured. Chemists agree that the rate of a reaction should be expressed as a positive number. As reaction 15–1–2 proceeds, the concentration of ozone becomes smaller while that of oxygen becomes larger. Thus the change in ozone concentration, Δ (conc O_3), is a negative quantity. To produce a positive rate in 15–1–3a, it is necessary to place the minus sign preceding the Δ (conc) term, as has been done.

In general, whenever the rate of a reaction is expressed in terms of the change in concentration of a reactant,D(conc reactant) is negative and the minus sign is necessary. The minus sign is unnecessary in 15–1–3b because Δ (conc O_2) is positive.

There is an additional difficulty, though. We would like to specify a single number representing the rate of a reaction under given conditions. However, the two expressions above give different numbers for the rate of 15–1–2, because dioxygen appears 1.5 times faster than ozone disappears (*i.e.*, in the same time that 2 moles of ozone react, 3 moles of dioxygen appear). This can be fixed by normalizing the two expressions in 15-1-3 with the stoichiometric coefficients. Dividing D(conc O_3) by the coefficient 2 gives the same number as dividing D (conc O_2) by the coefficient 3:

(15-1-4): Rate = –1/2 * Δ(conc O_3)/ time = 1/3 * D(conc O_2)/D time

Replacing "conc O_3" with the brackets to represent molar concentratration gives 15-1-5:

(15–1–5): Rate = –1/2 * Δ $[O_3]$/Dt = 1/3 * Δ $[O_2]$/D t

The symbol, Δ, signifies a finite (*i.e.*, measurable) change in a quantity. In the limit as the length of the time interval, Δt, approaches zero, we replace Δ(conc) and Δt with the differential (infinitesimal) quantities d(conc) and dt to obtain the *differential expression of rate* in 15-1-6.

(15–1–6): Rate = –1/2 Δ$[O_3]$/dt = 1/3 Δ$[O_2]$/dt

As we shall see in section 15–3, 15–1–5 is the appropriate formulation for finite difference methods of analysis.

For the completely general chemical reaction in 15–1–7, proceeding from left to right from reactants, A and B, to products, D and F, the rate of reaction can be represented using any one of the expressions in 15-1-8:

(15–1–7): aA + bB → dD + fF

(15–1–8): Rate = –1/a d[A]/dt = –1/b d[B]/dt = 1/d d[D]/dt = 1/f d[F]/dt

As discussed earlier, negative signs are necessary when rate is expressed in terms of disappearance of a reactant; and division by the stoichiometric coefficient guarantees that the same numerical value of rate is obtained from all four expressions.

Example. In the reaction of ozone to produce oxygen, it is found that under certain conditions of temperature and concentration, 0.0360 moles of ozone per liter react in a 2–hour period. How much dioxygen is produced in this time period? What is the average rate of reaction over this time period?

Solution. The chemical equation, 15–1–2, tells us that 3 moles of O_2 are produced for each 2 moles of ozone reacted.

Therefore 0.0360 moles ozone * (3 moles O_2/2 moles O_3) = 0.0540 moles dioxygen are produced.

The average rate of reaction can be expressed in either of two ways:

Rate = $-1/2\ \Delta[O_3]/\Delta t = -1/2\ (-0.0360$ moles/L)/2 hours = 0.009 moles/L–hour

Rate = $-1/3\ \Delta[O_2]/\Delta t = 1/3\ (0.0540$ moles/L)/2 hours = 0.009 moles/L-hour

Both expressions give the same positive rate of reaction.

Example: The rate of 15–1–9 was studied at 55°C by measuring the concentration of t–butyl bromide (t–BuBr) as a function of time. The data acquired are given below. Use the data to estimate the rate of reaction 20 seconds after reaction was begun.

(15–1–9): t-BuBr + $H_2O \rightarrow$ t-BuOH + HBr

Time, s	[t-BuBr]
0	0.100
10	0.0876
20	0.0768
30	0.0672
40	0.0590
50	0.0517
60	0.0453
80	0.0348
100	0.0267
120	0.0205
180	0.0093
240	0.0042

We begin by plotting the data to obtain a picture of the manner in which the concentration of t-BuBr varies with time. A plot of [t–BuBr] versus time is given in Figure. The shape of the plot is typical of a rate process: the concentration of reactant changes very rapidly at first, but as time goes on, the change in concentration per unit time becomes less and less as the concentration of t–BuBr approaches its final (equilibrium) value. We have seen this characteristic time plot before in our discussions of vapour pressure and approach to chemical equilibrium. Most concentration-time plots for a reaction in progress have this same general appearance.

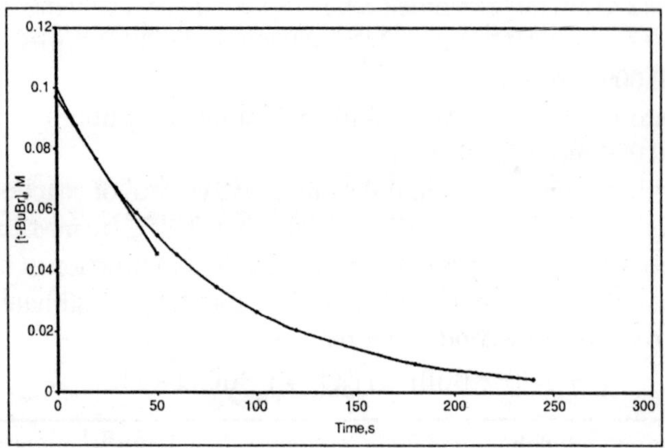

Fig. 10.1: Time Profile of The Concentration of T-Butyl Bromide

Two approaches are possible for obtaining the rate at the 20-second mark of the reaction. The finite difference expression of rate in 15–1–5 can be used to calculate the average rate during the time interval preceding or following the 20-second mark. Both average rates are calculated below:

Average rate between 10 and 20 seconds = –(conc at 20 s – conc at 10 s)/(20-10) Ms^{-1} = –(0.0768–0.0876)/(20-10) = 1.08 * 10^{-3} moles L^{-1} s^{-1}

Average rate between 20 and 30 seconds = –(0.0672–0.0768) (30-20) = 0.96 * 10^{-3} moles L^{-1} s^{-1}

The average rate over the 20–30 second time interval is less than that over the preceding interval because the reaction slows down as reactants are used up. This slowing down is readily apparent in the shape of the plot in Figure. However, the values are not too much different, so either may be used as an estimate of the rate of reaction at the 20 second mark.

The second approach is to use the differential expression of rate in 15-1-6. Just as dy/dx at $x = x_o$ is the slope of a plot of y versus x at $x = x_o$, d[t-BuBr]/dt at a particular time, t_o, is the slope of a plot of [t-BuBr] versus time at the particular time, t_o. This slope is the same as the slope of the tangent to the curve at t_o.

The tangent at t = 20 s is shown in the figure. It's slope is readily calculated to be 1.01×10^{-3} moles L^{-1} s^{-1}. As we might have expected, this value gives a rate intermediate between the values obtained above, based on finite time intervals. Because the slope of the tangent at a particular time represents the rate of reaction at that time, the rate obtained in this manner is called the *instantaneous rate* at the particular time. It is analogous to the speed that you measure when you glance at your automobile speedometer at a particular instant. As we have shown in this example, the instantaneous rate can be obtained by constructing tangents; or it can be estimated using the average rate over either the preceding or succeeding interval of time.

We can summarize our introductory discussion of reaction rate as follows:

- From a plot of [reactant] or [product] versus time, we can measure the rate at any particular time during the reaction as the slope of the tangent to the curve at that time; the initial rate is the slope of the tangent at t = 0.
- Rate decreases as the reaction proceeds towards equilibrium; this is evident from the shape of the concentration–time plot.
- The units of rate are []/time, or moles/L–time.

FACTORS AFFECTING REACTION RATE; THE RATE LAW

The rate of a chemical reaction depends on a number of factors, some of which are listed below:

- The molecular structure(s) of the reactant(s)
- The medium in which the reaction is carried out (for example, does the reaction occur in the gas phase, or does it occur with reactants dissolved in a solvent?)
- Temperature
- The presence or absence of a catalyst
- The concentrations of the reacting species

The dependence of rate on molecular structure is the most obvious but least understood factor. This complex matter is a focus of intense laboratory research. At present, it is not possible to summarize this dependence in a neat, clean package. In fact, this incompletely understood factor is what prevents predictive ability in kinetics. Consequently, we will postpone any discussion of this matter until a later chapter, in which the importance of chemistry in biology is discussed.

Similarly, the dependence of reaction rate on medium is a complicated matter that we will leave for more advanced books. Of the remaining three factors, temperature and the effects of catalysts are discussed in later sections of this chapter. In this section, we consider the effect of reactant concentrations on reaction rate.

For molecules to react, they must directly interact; to directly interact, they must collide. This leads us to expect that rate should be proportional to the frequency of molecular collisions. In turn, the frequency of collisions should depend on the number of molecules per unit volume, which is the concentration. This concentration dependence of rate is indeed observed experimentally. For most chemical reactions, the dependence of reaction rate on the concentrations of the reacting species is surprisingly simple.

In almost all cases, the reaction rate is found to be proportional to the product of the concentrations of the reactants, each concentration raised to an integer power that is almost always 1 or 2. Thus for the general reaction 15–1–7, the rate law is expected to take the form in 15–2–1.

$$(15\text{–}2\text{–}1): \text{Rate} = -1/a\ d[A]/dt = k\ [A]^x\ [B]^y$$

Equation-1 is called the *differential rate law* for 15–1–7, because it involves the ratio of differentials (derivative), d[A]/dt. The exponents, x and y, are integers, and for most reactions are either 1 or 2. It must be stressed that In most cases x and y are different from the stoichiometric coefficients, a and b; it is not valid to write the rate law for a reaction by raising the concentrations to the stoichiometric powers, unless the reaction is an *elementary process*. We will have more to say about elementary processes in a subsequent section. The integer, x, is called the *order of the reaction in reactant A*; y is the *order in B*.

The sum, x + y, is called the *overall reaction order*. The reaction orders must be determined by experiment; they can not be written down simply by looking at the reaction stoichiometry. To illustrate this statement for a specific case, the reaction below, involving 31 moles of reactants, has the accompanying rate law. This very complex overall reaction has a simple rate law, which is first order in each reactant, and second order overall.

$$30C_2H_5OH + B_{10}H_{14} \rightarrow 10B(OC_2H_5)_3 + 22H_2(g)$$

$$\text{Rate} = k[C_2H_5OH][B_{10}H_{14}]$$

Finally, k in 15–2–1 (and in the expression just above) is the proportionality constant relating Rate to the concentration product. It is called the *rate constant* for the reaction. The rate constant contains implicit within it information about the dependence of rate on molecular structure, reaction medium, and temperature. A word about the units of the rate constant is in order. Generally, k has whatever units are required to give the rate units of concentration/time. Example 15–3 shows how units for the rate constant may be determined for various rate laws.

Example. For each rate law below, state the order of the reaction in each reactant, and determine the units of the rate constant, k.

$$-d[A]/dt = k$$

$$-d[A]/dt = k[A]$$

$$-d[A]/dt = k[A]^2$$

$$-d[A]/dt = k[A][B]$$

$$-d[A]/dt = k[A]^2[B]$$

Solution.

Rate Law	Reaction Order	Units of k
$-d[A]/dt = k$	Zero order (no concentration dependence)	k has units of rate, Moles L^{-1} $time^{-1}$
$-d[A]/dt = k[A]$	First order in A	Moles/L-time = k(moles/L) so k has units of $time^{-1}$
$-d[A]/dt = k[A]^2$	Second order in A	Moles/L-time = $k(moles/L)^2$ so

		k has units L $moles^{-1}$ $time^{-1}$
$-d[A]/dt = k[A][B]$	First order in A First order in B Second order overall	Same as for the previous case
$-d[A]/dt = k[A]^2[B]$	Second order in A First order in B third order overall	Moles/L-time = $k(Moles/L)^3$ so k has units $L^2 moles^{-2}$ $time^{-1}$

SOME SPECIFIC EXAMPLES OF RATE LAWS. ZERO-ORDER REACTIONS

Chemical reactions having overall order zero are rare. However, evaporation of a pure liquid, discussed an example of a physical process with a zero-order rate law. Recall that the rate of evaporation of a liquid having exposed surface area, A, is just,

$$Rate = k^*A$$

As long as the exposed surface area remains constant, we may rewrite this as,

$$Rate = k', \text{ where } k' = k^*A$$

Since this rate is independent of concentration of liquid, the process is zero order in liquid.

FIRST-ORDER REACTIONS

Reactions and physical processes exhibiting first order rate laws are very common. *Condensation* is a first order physical process. The rate of condensation of vapour to liquid in a closed system is given by,

$$Rate = k^*P_{vap}$$

where P_{vap} is the vapour pressure of the substance. However, P is related to concentration in moles/L by the ideal gas law, leading to,

$$Rate = -d[vapour]/dt = k^*[vapour]$$

Molecular rearrangements (often called isomerization reactions) frequently follow first order rate laws. Vision is critically dependent upon, cis-retinal → trans-retinal

where cis- and trans- retinal contain the same atoms, bonded together in the same way, but with somewhat different spatial arrangements. The two forms are shown in Figure. The rate of conversion of cis- to trans- form is given.

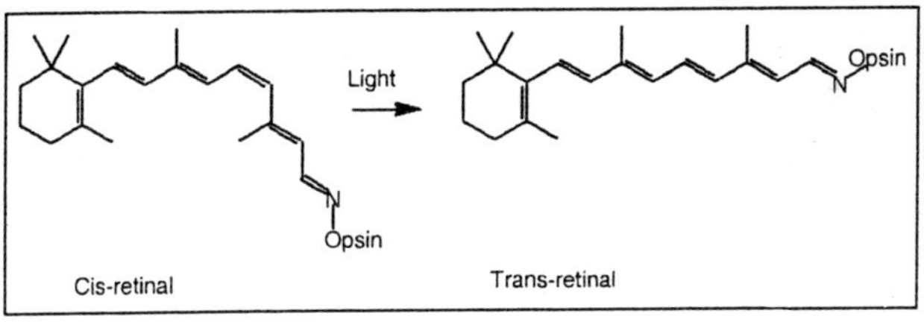

Fig. 10.2

$$-d[cis]/dt = k[cis]$$

Similarly, trans- converts to cis- by a first order rate law with a different rate constant. Finally, nuclear decay processes are first order. Thus ^{238}U decays according to 15-2-8, which is governed by the rate law in 15–2–9.

$$(15\text{–}2\text{–}8): {}^{238}U \rightarrow {}^{234}Th + {}^{4}He$$

$$(15\text{–}2\text{–}9): \text{Rate} = -d(\text{mass } {}^{238}U)/dt = k\ (\text{mass } {}^{238}U)$$

The rate constant for this process has the extremely small value, 1.54×10^{-10} year^{-1}. The rate constant, k, for a first-order process has units of time^{-1}. It's reciprocal, 1/k, thus has units of time. The quantity, 1/k, is given the symbol t, and is called the*time constant* for the reaction. This is a measure of the time required for the reaction to take place. The time constant for decay of ^{238}U is $1/k = 6.49 \times 10^9$ years. The fact that this time constant is so large is the reason why ^{238}U is still found in the earth's crust, which is estimated to be about 4.5 billion years old.

Example: The time constant for the radioactive isotope, ^{226}Ra, is 2338 years. Calculate the number of atoms of ^{226}Ra that decay each minute in a 2.00 g sample of ^{226}Ra.

Solution: Nuclear decay is a first order process. We can therefore write the following rate law for decay of ^{226}Ra:

Rate =–d(^{226}Ra atoms)/dt = k(Number of ^{226}Ra atoms)

$$-dN/dt = kN$$

where N is the number of ^{226}Ra atoms in the sample, and $k = 1/\tau = 4.28 \times 10^{-4}$ year^{-1}.

$$N = (2.00g/226.025\ g/mole) * 6.023 * 10^{23}\ atoms/mole = 5.33 * 10^{21}\ atoms.$$

$$-dN/dt = (4.28 * 10^{-4}\ year^{-1})(5.33 * 10^{21}\ atoms) = 2.28 * 10^{18}\ atoms/year$$

There are $5.256 * 10^5$ minutes in 1 year. Therefore,

$$-dN/dt = (2.28 * 10^{18} atoms/year)/(5.256 * 10^5\ minutes/year) = 4.34 * 10^{12}\ atoms/minute$$

SECOND ORDER PROCESSES

Many reactions exhibit second-order rate laws, either *simple* (second-order in a single reactant) or *mixed* (first-order in each of two reactants). Here we consider two specific reaction types that exhibit second-order kinetics. *Dimerization reactions,* in which two molecules unite to form a larger molecule, obey simple second-order kinetics. The dimerization of NO_2 is shown in 15–2–10.

$$(15–2–10): 2NO_2(g) \rightarrow N_2O_4(g)$$

The rate law for 15–2–10 is 15-2-11:

$$(15–2–11): -d[NO_2]/dt = k[NO_2]^2$$

where the second order rate constant is—$M^{-1}s^{-1}$ at °C. Other examples of dimerization processes are the formation of a Cl_2 molecule from two Cl atoms and the formation of one molecule of C_4H_8 from two molecules of C_2H_4. In general, a reaction of the general type in 15–2–12 is expected to obey simple second-order kinetics.

$$(15–2–12): 2\ A \rightarrow A_2$$

Adduct formation processes, such as Lewis acid-base reactions and the formation of complexes involving enzymes and their substrates, are also second-order processes. Most such processes occur very rapidly. The role of enzyme-substrate complexes in enzyme catalysis will be discussed in a subsequent section.

"NET" REACTION RATE; A PRELUDE TO THE NEXT SECTION

In the next section we discuss experimental approaches to the determination of rate laws for reactions. To preface that discussion, we briefly discuss here the concept of *net reaction rate*. Except in some cases, we can experimentally measure only the net rate of a reaction, where the net rate is defined as in 15–2–13.

(15–2–13): Net reaction rate = rate of forward reaction – rate of reverse reaction

$$Rate_{net} = Rate_{forward} - Rate_{reverse}$$

The general reaction is reproduced below, this time acknowledging that in general the reverse process also occurs:

(15–1–7): $aA + bB \leftrightarrow dD + fF$

For this reaction we expect the forward rate to depend only on the concentrations of reactants, A and B, and the reverse rate to depend only on the concentrations of products, D and F, as shown in 15–2–14:

(15–2–14a): $Rate_{forward} = 1\ k_f[A]^x[B]^y$

(15–2–14b): $Rate_{reverse} = k_r[D]^p[F]^q$

The net rate is therefore expected to be a potentially complex function of the concentrations of reactants AND products. The experimentally determined rate constant is a composite function of the forward and reverse rate constants, k_f and k_r.

(15–2–15): $Rate_{net} = k_f[A]^x[B]^y - k_r[D]^p[F]^q$

Once reaction reaches equilibrium, the net rate will of course be zero. Applying this idea to 15–2–13 leads to 15–2–16, our familiar criterion for dynamic equilibrium:

(15–2–16a): $Rate_{net} = 0 = Rate_f - Rate_r$

(15–2–16b): $Rate_f = Rate_r$

There are a few situations in which the net rate is equal to the forward reaction rate:

- One of the products escapes from the reaction system, so that its concentration remains at or very near zero at all times. A gaseous product that escapes from a solution is an example of this situation.

- The reaction has a large value of K_{eq}, so that the reverse rate is negligible except near equilibrium, which is not achieved until almost all of the limiting reactant has been used.
- At the very beginning of a reaction in which only reactants are initially present. At time zero, the reverse rate must be zero because no products are present yet.

EXPERIMENTAL DETERMINATION OF THE RATE LAW. THE INITIAL RATE METHOD

When applicable, the method of initial rates is a simple and clean way to obtain the rate law. The initial rate of a reaction is the instantaneous rate at t=0, when reactants are first mixed together and begin to react. In a plot of reactant concentration versus time, it is given by the slope of the tangent at t = 0. The slope of this type of plot is always steepest at the beginning; the initial rate is thus the fastest rate that we observe during the time course of the reaction process. One major advantage of the initial rate method is that the reverse rate is zero, provided only reactants and NO products are initially present. Thus only the forward rate contributes to the initial rate, and complications resulting from the reverse reaction are avoided. *The initial rate law depends only on concentrations of reactants.* The method of initial rates is best illustrated by example.

Example. The initial rate was measured at several combinations of initial reactant concentration for the following reaction:

$$A + 2B + 3C \rightarrow D + E$$

The rate law is expected to have the following general form:

$$d[A]/dt = k[A]^x[B]^y[C]^z$$

The following data were obtained. Use the data to determine the values of x, y and z in 15–3–2.

Run Number	$[A]_o$, M	$[B]_o$, M	$[C]_o$, M	Initial Rate, M/s
1	0.05	0.10	0.10	6.00×10^{-3}
2	0.05	0.15	0.10	13.5×10^{-3}
3	0.05	0.15	0.05	13.5×10^{-3}
4	0.10	0.10	0.05	12.0×10^{-3}

Solution. The key to the initial rate method is to *examine the dependence of rate on one reactant at a time, using data obtained when the other reactants do not vary in initial concentration.* In the first two data entries above, the initial concentrations of both A and C are the same in both runs; only the initial concentration of B is changed. These two data items can therefore be used to determine the order of the reaction in B. We see that when the initial concentration of B is increased from 0.10 to 0.15, an increase of a factor of 1.5, the rate increases from 6 to 13.5×10^{-3} moles/L-s, which is a factor of 2.25. We translate this information into equation form:

$$\text{Rate 2/Rate 1} = [B]_2^y/[B]_1^y = ([B]_2/[B]_1)^y \; 2.25 = (1.5)^y$$

We recognize that y must be 2. The reaction is therefore second order in B, and the rate law thus far is,

$$-d[A]/dt = k[A]^x[C]^z[B]^2$$

In runs 2 and 3, the initial concentrations of A and B are the same, while that of C is changed. We see from these data that even though the concentration of C is only half as large in run 3 as in run 2, the rate is unaffected. We conclude that the rate does not depend on [C], hence that $z = 0$. The rate law to this point is,

$$-d[A]/dt = k[A]^x[B]^2$$

There are no two kinetics runs in which only [A] is varied. However, runs 1 and 4 can be used to obtain the order in A because we have just determined that rate does not depend on [C]. Runs 1 and 4 indicate that the rate doubles when the initial concentration of A doubles. The order in A is thus 1, and the final rate law is,

$$-d[A]/dt = k[A][B]^2$$

Once the rate law is known, the rate constant may be calculated from any one of the individual runs. Choosing the first run,

$$k = \text{Rate}/[A][B]^2 = 6.00 * 10^{-3}/(0.05)(0.10)^2 = 12.0 \text{ L}^2/\text{mole}^2\text{s}$$

As this example shows, the initial rate method, *when it is applicable,* provides a clean, simple path to the rate law. The italicized qualifier is important, however; in practice, the method of initial rates is only rarely used, for several reasons. In most kinetics studies, the chemist examines the manner in which the concentration of a selected reagent varies with time. The resulting data are presented either in a table, or in a plot of concentration versus time, as in Example and Figure. The data obtained at the beginning of the reaction, where the rate is largest and the time interval is smallest, is usually less reliable than that obtained at later times, when the rate is less. It is undesirable to base an analysis on one's least reliable data. Second, and more important, the method of initial rates uses only a small fraction of the concentration time data; the rest is wasted.

Third, to apply the initial rate method as was done in the example above requires that full concentration-time profiles be collected for each run, in order to extract the slope of the concentration-time plot at t=0.

Collection of this much data requires a lot of experimental effort. Most of the resulting data is then ignored if the initial rate method is used. For these reasons and others, chemists have developed other methods for determining rate laws from data, which make use of all of the data collected. Some of these will be discussed now.

INTEGRATION OF THE RATE LAW

It is possible to convert the differential rate law into an integrated form, in which concentration is expressed directly as a function of time. The concentration-time data obtained during a kinetics run in the lab may then be matched against the function. If the data conform to the function, it may be concluded that the reaction follows the corresponding differential rate law. Integration is very easy for rate laws of the form

$$-d[A]/dt = k[A]^n$$

We will examine the first and second–order cases, with n = 1 and 2 respectively. The first order rate law may be rearranged to put the concentration terms on one side and time on the other side of the equals sign, as in,

$$-d[A]/[A] = kdt$$

Both sides of the equation are then integrated over the limits, $[A] = [A]_o$ at t=0 and $[A] = [A]_t$ at some later time t:

$$\int_{[A]0}^{[A]t} d[A]/[A] = k\int_0^t dt$$

The integral of dx/x is ln x, and the integral of dx is x. Evaluating over the limits gives:

$$\ln ([A]_t/[A]_o) = -k\,t$$

This is the so-called *integrated first order rate law*. For a reaction following simple first order kinetics, a plot of ln $[A]_t$ versus t is linear, with slope –k. Thus if concentration-time data for a reaction are plotted according to 15–3–9, and linearity is obtained, it may be concluded that the reaction follows a simple first-order differential rate law. An alternative expression of 15–3–9 is obtained by exponentiating both sides:

$$[A]_t = [A]_o e^{-kt}$$

Here e = 2.718... is the base of natural logarithms. Substituting the reciprocal of the time constant, t, for the rate constant gives,

$$[A]_t = [A]_o e^{-t/\tau}$$

Thus when an interval of time equal to the time constant has elapsed, the concentration of A will have decreased to 1/e (about 1/3) of its initial value. With the integrated first-order rate law in hand, we are in a position to consider a quantity called the *half-life*, $t_{1/2}$, which is similar but not identical to the time constant.

The half-life for a process is defined as the time required for the concentration of the limiting reactant to decrease to 1/2 of its initial value. An expression for the half life is easily obtained from 15–3–9 by realizing that $[A]_t = [A]_o/2$ at $t = t_{1/2}$. Making these substitutions in 15-3-9 leads to the following expression for the half-life of a first-order reaction:

$$t_{1/2} = \ln 2/k = 0.693/k$$

For n = 2 in corresponding to a second-order process, rearrangement leads to:

$$-d[A]/[A]^2 = k\ dt$$

Integration of both sides of this equation results in 15–3–11:

$$1[A]-1/[A]_o = k\ t$$

For a reaction with a simple second order rate law, a plot of concentration-time data as 1/[A] versus t is linear, with slope k. Thus a reaction for which the concentration-time data produce a linear 1/[A] versus t plot may be concluded to follow a second-order rate law.

Example: Plot the concentration-time data in Example according to equations and to determine whether the hydrolysis of t-butyl bromide obeys a simple first- or second-order rate law.

Solution: The concentration-time data from Example are reproduced below. Columns have been added for ln [t–BuBr] and 1/[t–BuBr]

Time, s	[t–BuBr], moles/L	ln [t–BuBr]	1/[tBuBr]
0	0.100	–2.303	10
10	0.0876	–2.435	11.42
20	0.0768	–2.567	13.02
30	0.0672	–2.700	14.88
40	0.059	–2.830	16.95
50	0.0517	–2.962	19.34
60	0.0453	–3.094	22.08
80	0.0348	–3.358	28.74
100	0.0267	–3.623	37.45
120	0.0205	–3.887	48.78
180	0.0093	–4.678	107.53
240	0.0042	–5.473	238.10

Figure is a plot of ln[t–BuBr] versus time; and Figure shows 1/[t–BuBr] against time. The linearity of the first plot indicates that the reaction is first–order in t–BuBr. The rate constant, obtained from the slope of Figure, is $1.32 * 10^{-2}\ s^{-1}$ (recall that for a first order reaction, the units of the rate constant must be $time^{-1}$, resulting from the division of rate in conc/time by concentration).

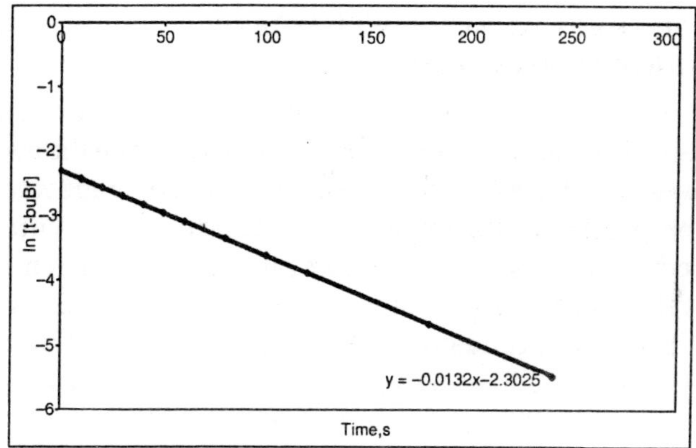

Fig. 10-3a: In [T-Bubr] Versus Time

Example: One product resulting from fission of ^{235}U in nuclear reactors is ^{137}Cs, a radioactive nucleus with a time constant of 43.3 years. How much time will it take for the radioactivity level of this waste to fall to 1% of its initial value, which is considered a safe level?

Solution: This problem is readily solved using equation 15-3-9:

$$\ln [A]/[A]_o = -kt = -t/\tau$$

$$t = -\ln(0.01)\tau = -\ln(0.01)(43.3) = 200 \text{ years}$$

The waste must be safely stored for 200 years before it will be safe.

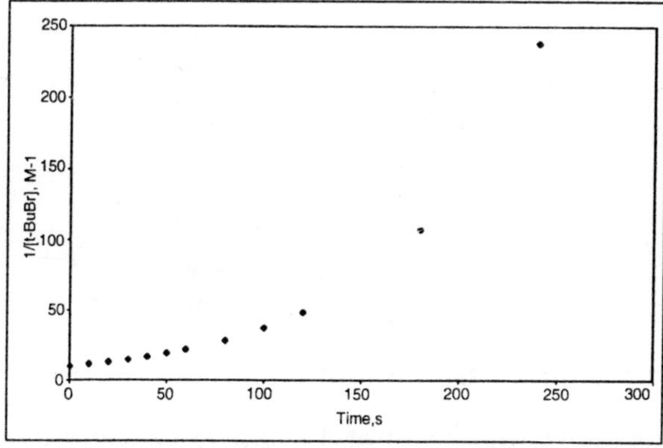

Fig. 10-3b: 1/[T-Bubr Vs. Time

FINITE DIFFERENCE METHODS OF NUMERICAL ANALYSIS

Another approach to determination of the rate law and rate constant from concentration-time data is to calculate the average rate over each experimental time interval. The average rates can then be plotted in turn against concentration, concentration squared, and so on until a linear plot is obtained. This is a relatively simple way to determine whether a reaction is first-order or simple second order (that is, has a rate law of the form, Rate = $k[A]^2$). Estimating derivatives (slopes) and integrals (areas) using finite changes in variables is called the Method of Finite Differences. We illustrate this method using the data for hydrolysis of t-BuBr from Examples 15–2 and 15–6.

Example: Use finite difference methods to obtain the rate law and rate constant for the hydrolysis of t-butyl bromide according.

Solution: The time and concentration data from Example 15–2 are reproduced below. In addition, values of Δt, Δ[t–BuBr], and rate = Δ[t–BuBr]/Δt have been calculated for each pair of successive data points.

Time, s	Δt, s	[t–BuBr], moles/L	Δ[t–BuBr], moles/L	Rate * 10^3
0		0.100		
10	10	0.0876	–.0124	1.24
20	10	0.0768	–.0108	1.08
30	10	0.0672	–.0096	0.96
40	10	0.0590	–.0082	0.82
50	10	0.0517	–.0073	0.73
60	10	0.0453	–.0064	0.64
80	20	0.0348	–.0105	0.525
100	20	0.0267	–.0081	0.40
120	20	0.0205	–.0062	0.31
180	60	0.0093	–.0112	0.19
240	60	0.0042	–.0051	0.085

Thus the change in t-butyl bromide concentration over the first 10-second interval is the concentration at the 10 second mark minus the initial concentration at t = 0. The average rate over this interval is calculated by dividing the concentration change by the duration of the interval, Δt. Figure 15–4 shows plots of Rate versus [t-BuBr] in a and Rate versus $[\text{t-BuBr}]^2$ in b.

The linearity of the plot in Figure confirms that the reaction is first order in t-butyl bromide. The slope of the plot gives k = $1.36 * 10^{-2}\ s^{-1}$. This value is in satisfactory agreement with the value obtained from the ln[t-BuBr] versus time plot in Example 15-6. Even better agreement can be achieved by plotting average rate over a time interval versus the average [t-BuBr] over the same time interval.

When the above data are plotted in this manner, the rate constant is found to be $1.33 * 10^{-2}\ s^{-1}$. Note that the method of finite differences, like the method based on integrated rate laws, uses the full range of concentration–time data, thus avoiding the major problem with the method of initial rates. For the method to be successful, however, it is important that the time intervals over which average rates are calculated be short relative to the total reaction time.

In the example above, the 10-second time intervals used in the early stages of the reaction, when concentration changes most rapidly, are small compared with the overall reaction time of 240 seconds. As a useful guideline, time intervals should not exceed 5% of the total reaction time; otherwise inaccurate rate constants will result from the finite difference approach.

Electronic spreadsheets enable the calculations required for the finite difference method to be performed rapidly and conveniently using a computer. Further, the various data plots can be made quickly and professionally using the spreadsheet. With this modern tool, calculation by finite difference methods is routine. We will consider one additional example of the approach.

Example: The decomposition of N_2O occurs according to the following equation:

$$2N_2O(g) \rightarrow 2N_2(g) + O_2(g)$$

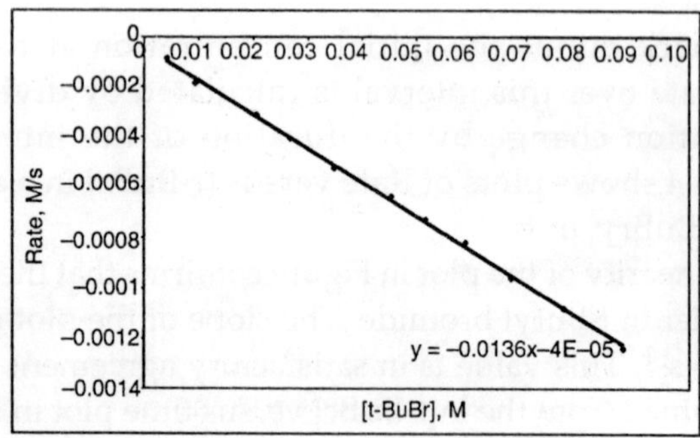

Fig. 10-4a: Rate Vs. [T-Bubr]

In a kinetics run, the concentration of N_2O was found to change with time as indicated in the first 2 columns of Table. The numbers in the remaining columns, calculated using an electronic spreadsheet, represent the average $[N_2O]$ over each time interval, squares of the N_2O concentrations, and the average rates of N_2O concentration change over the successive time intervals. Figure 10–4a and b show spreadsheet–generated plots of Rate versus $[N_2O]$ and Rate versus $[N_2O]^2$, respectively. Plot a shows slight but distinct curvature and in addition has a non–zero intercept. The linearity of the plot and lack of intercept in b confirms that the reaction follows the rate law in 15–3–13:

$$-d[N_2O]/dt = k[N_2O]^2$$

The rate constant, obtained from the slope of the plot, is $6.0 * 10^{-3}$ L/mole-min.

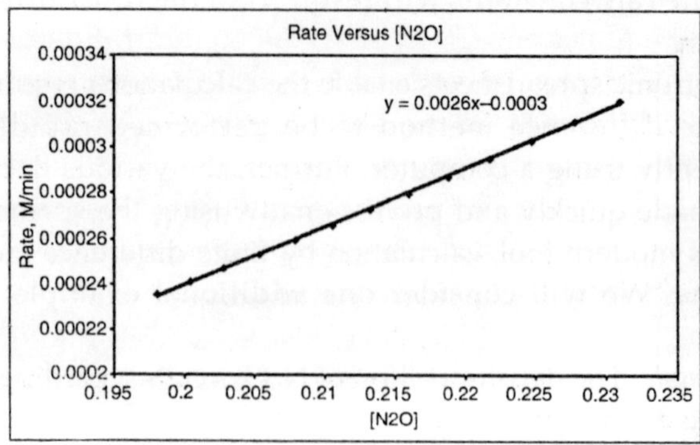

Fig. 10.5

PSEUDO-ORDER METHODS

Many chemical reactions obey a mixed second-order rate law of the type shown:

$$\text{Rate} = k[A][B]$$

The reaction is first order in each of two reactants. An example of a reaction governed by this rate law is given in 15–3-15.

$$NH_4^+ + NO_2^- \rightarrow N_2(g) + 2H_2O$$

Suppose that a kinetics experiment is carried out by preparing a solution containing similar initial concentrations of the two reactants, NH_4^+ and NO_2^-, and monitoring the appearance of $N_2(g)$ as a function of time in order to measure the rate. Since the two reactants are present in similar concentrations, both substances undergo significant decreases in concentration over the course of the reaction. The rate of reaction reflects this simultaneous decrease in concentration of both reactants. Consequently, a plot of Rate versus $[NH_4^+]$ (NH_4^+ = A) will be non-linear, even though the reaction is first order in NH_4^+, because the Rate also responds to the change in concentration of NO_2^- (B) significantly during reaction. Although it is possible to integrate 15–3–14 to obtain a rather complex equation that may then be used to plot the data, a simpler approach to the kinetics study is often possible.

In this approach, a kinetics run is performed on a solution that initially contains a substantial excess of one of the reactants, say B. If the initial concentration of B is more than 10 times larger than the initial concentration of A, the concentration of B will be nearly the same at the end of the reaction as at the beginning; consequently it may be considered to be constant during reaction. To illustrate this, consider a solution initially containing 0.002 M NH_4^+ and 0.10 M NO_2^-. If reaction is allowed to proceed until all NH_4^+ is consumed, the concentration of NO_2^- will have decreased by only 0.002 M to 0.098 M:

	NH_4^+ +	$NO_2^- \rightarrow$	N_2 +	$2H_2O$
Initial	0.002	0.10	0	
Change	–0.002	–0.002	0.002	
Final	0	0.098	0.002	

Thus the concentration of NO_2- remains essentially constant while NH_4^+ is consumed, and the rate law simplifies as shown in 15–3–16:

$$d[N_2]/dt = k[NH_4^+][NO_2^-] = k'[NH_4^+]$$

Here the constancy of $[NO_2^-]$ has been recognized, and this concentration has been lumped into the rate constant to produce a new constant k'. The rate law now has a simple first-order form, even though the reaction is actually second-order. This simplified rate law, which may be integrated or subjected to finite difference analysis, is called a *pseudo-first order* rate law.

The prefix, pseudo-, reminds us that the reaction is not truly first-order, but has only been made to appear that way by imposition of a large concentration of NO_2^-. If one carries out kinetic runs using several (relatively large) concentrations of [B], a pseudo-first order rate constant, k', will be obtained for each run. Of course, these will have different values depending on [B]. A plot of k' versus [B] is then made to obtain the true rate constant, k, from the slope.

Example. Pseudo-First Order Kinetics. For the reaction of NH_4^+ with NO_2^-, 15–3–15, several kinetics runs were carried out using the initial reactant concentrations in the first two columns of the table below. Plots of Rate versus $[NH_4^+]$ for each run were found to be linear, indicating that the reaction is first-order in NH_4^+. The first-order rate constants obtained from these plots varied with $[NO_2^-]$ as indicated in the third column of the table. What is the order of the reaction in NO_2^-? What is the true rate constant for the reaction?

$[NH_4^+]$Initial	$[NO_2^-]$Initial	Observed 1st Order Rate Constant, k'
0.00896	0.197	$7.06 * 10^{-5}$
0.00904	0.395	$1.42 * 10^{-4}$
0.00523	0.593	$2.13 * 10^{-4}$

Solution. The order of reaction with respect to NH_4^+ is known to be 1, based on linear Rate vs. $[NH_4^+]$ plots. We thus focus on the relationship between the observed pseudo-first-order rate constants and the concentration of $[NO_2^-]$. Note that for all three runs, the initial NO_2^- concentration is much larger than the initial $[NH_4^+]$, guaranteeing that $[NO_2^-]$ remains essentially constant during the entire course of reaction.

A linear plot of k′ versus $[NO_2^-]$ raised to a power, n, indicates that the reaction is n-order in $[NO_2^-]$. We can determine n by trying several plots, first with n = 1, then, if necessary, with n = 2, 3, and so on, until we obtain linearity. Plots of k′ versus $[NO_2^-]$ and $[NO_2-]^2$ are shown in Figure. The top plot is linear, so the reaction is first order in $[NO_2^-]$. The overall second-order rate constant, $k = 3.6 * 10^{-4}$ L/mole-s, is obtained from the slope.

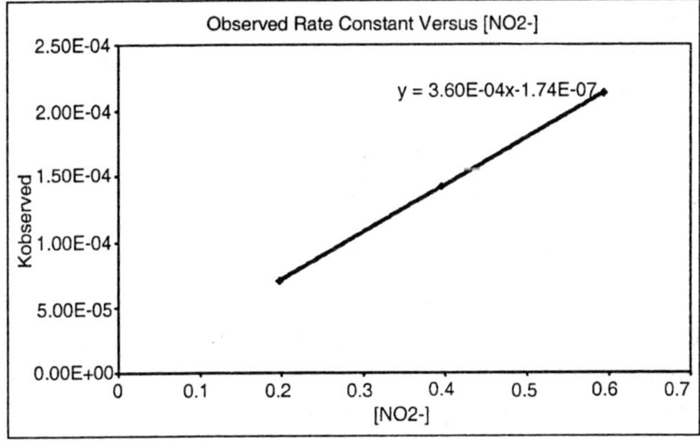

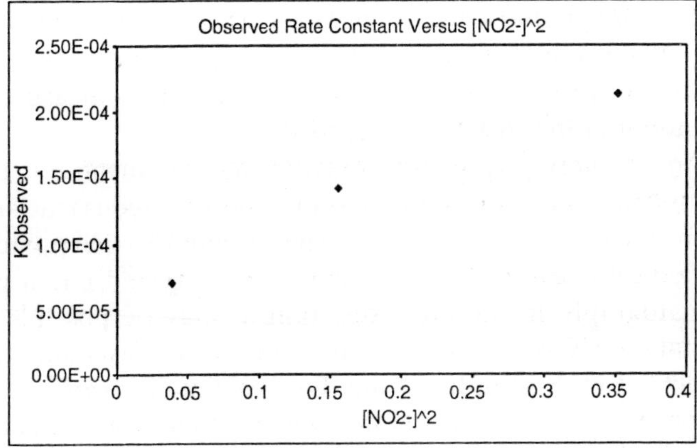

Fig. 10.6

THE EXPERIMENTAL MEASUREMENT OF CONCENTRATION

In order to carry out a kinetics study, it is necessary to have an experimental "handle" on at least one of the reactants or products; that is, the change in concentration of a reactant or product must be accompanied by a change in some measurable property of the system. For reactions involving gases, the total pressure of the reaction mixture changes as reaction proceeds, provided that there is a net change in the amount of gas in the system. Thus, the rate of reaction 15-3-17 could be studied by monitoring the total pressure with time. Total pressure decreases as reaction proceeds because only 1 mole of gas forms for each 2 moles consumed.

$$2C_4H_6 \rightarrow C_8H_{12}$$

On the other hand, no change in total pressure accompanies 15-3-18 because there is no change in moles of gas; reaction rate can not be studied by monitoring total pressure in this case.

$$H_2(g) + I_2(g) \rightarrow 2HI(g)$$

Rather than measuring system pressure at constant volume, one could measure system volume at constant pressure. In either case, the measured quantity is proportional to the number of moles of gas in the system at a particular time during reaction. Pressure (or volume) measurements are particularly convenient when only a single gaseous reactant or product is involved. In this case, the pressure (or volume) provides a direct measure of the amount of that reactant or product.

An alternative to pressure or volume measurements is to use gas chromatography to monitor the amounts of gaseous reactants and products as a function of time. The gaseous reaction mixture is periodically sampled with a syringe and injected into a gas chromatograph. In the ideal situation, it may be possible to simultaneously determine the amounts of all reactants and products at each sampling time. For reactions carried out in solution, a number of methods are used to monitor concentration. For 15–3–19, the acidity of the solution changes during the course of reaction. Provided the reaction is slow enough, the kineticist could periodically "sample" an aliquot of reaction solution and carry out an acid-base titration using a suitable indicator to determine the $[H_3O^+]$ present at the known sampling time.

$$t\text{–}BuCl + 2H_2O \rightarrow t\text{–}BuOH + H_3O^+ + Cl^-$$

Alternately, since ions are produced but not consumed in the reaction, the kinetics could be monitored by measuring the electrical conductivity of the solution as a function of time. The measured conductivity values can then be related to the total concentration of ions present at that time. Perhaps the most common method for following the time course of reactions in solution is to monitor the absorption of light by the reacting solution. If one of the reactants (or products) has an absorption band in the UV–visible, infrared, or NMR region of the electromagnetic spectrum, the change with time of the amount of light absorbed is a measure of the rate of appearance or disappearance of the absorbing substance. The kinetics of 15–3–20 can easily be studied by this method.

$$Cr(H_2O)_6^{3+} + EDTA^{4-} \rightarrow Cr(EDTA)^- + 6H_2O$$

The product, $Cr(EDTA)^-$, has an absorption maximum at a wavelength of 542 nm in the visible region of the spectrum. As reaction proceeds, absorbance at this wavelength increases in direct proportion to the concentration of $Cr(EDTA)^-$, providing a "handle" on the reaction kinetics.

ELEMENTARY PROCESSES AND REACTION MECHANISM

A balanced chemical equation is intended to represent the *net* transformation of reactants to products that occurs when reaction takes place. It does not, and is not intended to, tell us anything about the details of the reaction pathway. In fact, most reactions occur via a series of relatively simple steps, each involving reaction of only a few (one, two, and very occasionally, three) molecules. Each of these simple steps is called an *elementary process*. The equation for an elementary process represents an actual collision event at the molecular level. The sequence of elementary processes by which a reaction is proposed to occur is called the *mechanism* for the reaction. For example, the reaction in 15–4–1 is thought to occur by the 2-step mechanism in 15–4–2a and 15–4–2b.

Overall:

$$(15\text{–}4\text{–}1): 2NO_2(g) + F_2(g) \rightarrow 2NO_2F(g)$$

Mechanism:

$$(15\text{–}4\text{–}2a): NO_2 + F_2 \rightarrow NO_2F + F$$

$$(15\text{–}4\text{–}2b): NO_2 + F \rightarrow NO_2F$$

with the first step proposed to be slow, the second one fast. Each of the two steps in the mechanism is understood to be an elementary process. We can make several important generalizations about the mechanism in 15–4–2. First, each elementary process involves reaction (collision) of only two species, even though the overall reaction involves 3 moles of reactants. Second, F atoms, produced in the first step and consumed in the second, do not appear in the overall equation for the reaction. Such a species—first produced, then consumed—is called a *reaction intermediate*. Third, the elementary processes add to give the overall equation for the reaction. This must always be true. It is important to realise that a mechanism for a reaction is a hypothesis about what goes on at the molecular level, made by the chemist based on her or his experimental observations. The overall equation for a reaction is NOT a hypothesis; it is an experimental fact, observable in the laboratory. The mechanism in all of its detail is seldom observable; it must be deduced from experiment, must be consistent with all known experimental facts about the reaction, and is subject to change if in conflict with a new experimental fact.

TYPES OF ELEMENTARY PROCESSES

There are three types of elementary process, named for their *molecularity*—the number of reactant molecules that they involve. A *unimolecular process* involves only one reactant molecule. Two examples of unimolecular processes are given.

$$O_3 \rightarrow O_2 + O$$

$$I_2 \rightarrow 2I$$

A *bimolecular process* involves two reactant molecules, which may be the same or different. Examples are given. Bimolecular elementary processes are by far the most common type.

$$NO_2 + F \rightarrow NO_2F$$

$$H_2O_2 + I^- \rightarrow H_2O + IO^-$$

A *termolecular process* requires the simultaneous collision of three reactant molecules. is an example.

$$2NO + O_2 \rightarrow N_2O_4$$

Such three-body collisions are extremely improbable. Consequently, termolecular processes are rare. Elementary processes with molecularity greater than three do not occur, due to the overwhelming improbability of collisions involving four or more molecules. In contrast to the situation for the overall (net) reaction, *the rate law for an elementary process may be written directly from its stoichiometry,* because an elementary process is an actual, not a net, process. Thus for the elementary process,

$$A + B \rightarrow C + D$$

the rate law is $-d[A]/dt = k[A][B]$. Unimolecular processes have first-order rate laws. For, the rate law is

$$-d[O_3]/dt = k[O_3]$$

Bimolecular processes have rate laws that are second-order overall; and termolecular processes have overall third order rate laws.

$$-d[NO_2]/dt = k[NO_2][F]$$

$$-d[O_2]/dt = k[NO]^2[O_2]$$

In summary, an elementary process of molecularity n has a rate law of order n. The converse is NOT true: a first-order reaction is not necessarily either an elementary process or unimolecular. The difference between order and molecularity is subtle but important. The concept of molecularity is appropriate only in the context of an elementary process, and indicates the number of reactant molecules involved in (undergoing collision) in the elementary step. The order of a reaction is the number to which the concentration of a reactant must be raised in the experimental rate law for the overall (net) reaction. This rate law, and the orders in it, do not necessarily correspond to a particular elementary step in the mechanism of the net reaction. We will say more about this shortly.

RATE DETERMINING STEP

A particular overall reaction consists of a sequence of elementary steps, which usually occur with different rates. Very often one of the elementary processes occurs much more slowly than do the others and limits the rate at which the overall reaction occurs. This slow step is then referred to as the*rate-determining step (rds). The overall rate is equal to the rate of the rds*. The fact that the overall reaction rate is limited by the rate of a particular elementary step is the reason why rate laws frequently do not correspond with the stoichiometry of the overall reaction. A familiar example of an overall process occuring at a rate limited by a slow step is illustrated in Figure. Pictured is a theater filled to capacity, connected by a set of wide double doors to a smaller lobby, which is linked to the outside via a revolving door. When the movie is over, people begin to move from the theater to the lobby, and from there to the outside. We suppose that the double doors between theater and lobby are large enough so that 10 people per second can pass through; but that only 1 person per second can exit the revolving door. We then represent the overall process of leaving the theater in terms of 2 steps:

theater $\rightarrow$ lobby [10/s]

lobby $\rightarrow$ outside [1/s]

Overall, theater $\rightarrow$ outside [Rate = ?]

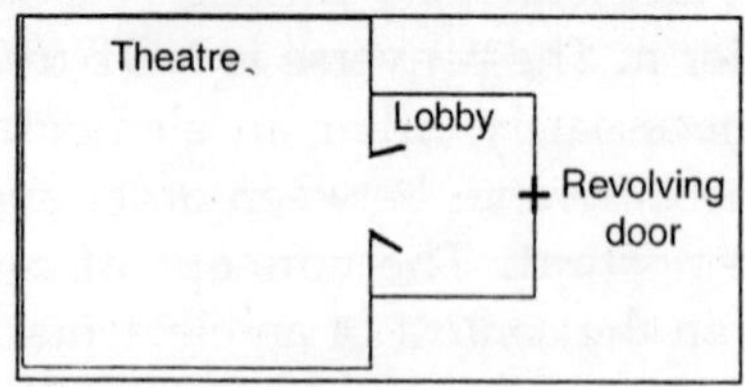

Fig. 10.7

Initially, people pass from theater to lobby at the rate of 10 per second, and begin to exit the revolving door one by one.

However, the lobby quickly fills up, because it cannot be emptied nearly as rapidly as it was filled. The process "piles up" behind the slow step, which is passage through the revolving door. The maximum rate at which people can leave the theater is 1 per second, which is the rate of the slowest step in the process.

RATE LAW AND REACTION MECHANISM

Chemists are very interested in the details of how a reaction occurs; *i.e.*, in the reaction mechanism.

The approach to the elucidation of the mechanism of a reaction of known stoichiometry involves at least the following steps:

- Do experiments to determine the experimental rate law for the overall reaction;
- Propose a mechanism for the overall reaction, based on the observed rate law;
- Derive the theoretical rate law implied by the proposed mechanism;
- Compare the rate laws in 1 and 3. If they have the same form, it is safe to conclude that the mechanism is consistent with the experimental rate law. If they do not have the same form, go back to 2.

Step 2 makes use of the fact that the *observed rate law indicates what atoms are involved as reactants in the rate-determining step of the mechanism*. This will be made clear in specific examples to follow. It is the most difficult step of the above procedure, requiring extensive experience and well-developed "chemical intuition." At this level of study, we will not attempt to deduce mechanism from the rate law. Rather, our focus will be this: given the form of the experimental rate law, and one or more proposed mechanisms, decide whether each mechanism is or is not consistent with experiment.

Example: For the reaction $2NO_2(g) + F_2(g) \rightarrow 2NO_2F(g)$, the experimental rate law has the form, Rate = $k_{obs}[NO_2][F_2]$. Two mechanisms have been proposed. These are given below:

Mechanism 1

$$NO_2 + F_2 \Leftrightarrow NO_2F + F \; [k_1, k_{-1} \text{ fast}]$$
$$NO_2 + F \rightarrow NO_2F \; [k_2 \text{ slow (rds)}]$$

Mechanism 2

$$NO_2 + F_2 \rightarrow NO_2F + F \; [k_1 \text{ slow (rds)}]$$

$$NO_2 + F \rightarrow NO_2F \; [k_2 \text{ fast}]$$

Can either of these mechanisms be ruled out as being in conflict with experiment?

Solution: We examine mechanism 1 first. To obtain the theoretical rate law for this mechanism, we proceed as follows:

1. Write the rate law for the rds. The rate of this step is the same as the rate of the overall reaction.

The rds for mechanism 1 is step 2.

$$\text{Rate} = k_2[NO_2][F]$$

Although this rate law is valid, it is not expressed solely in terms of concentrations of the reactants in the overall reaction. Since these concentrations are the ones we actually observe, we must somehow rewrite the rate law so as to eliminate the concentration of the intermediate, F, which is not generally observable, and replace it with concentrations of reactants in the net reaction. We can accomplish this making use of the preceding rapid steps, which we assume to occur rapidly enough to come to a pseudo equilibrium. So,

2. Use the rapid pre-equilibrium to solve for the concentration of the intermediate in terms of known concentrations. Since the first step of the mechanism is in equilibrium, the forward and reverse rates of this elementary process are the same, and we can write

$$k_1[NO_2][F_2] = k_{-1}]NO_2F][F]$$

Solving for [F] gives,

$$[F] = k_1/k_{-1} \{[NO_2][F_2]/[NO_2F]\}$$

This expression for the concentration of the intermediate can be substituted into the rate law for the rds:

$$\text{Rate} = k_1k_2/k_{-1} \{[NO_2]^2[F_2]/[NO_2F]\}$$

This is the rate law implied by the mechanism. It is the rate law that the reaction should follow if the proposed mechanism is correct. Clearly, though, it is not in agreement with the experimental rate law, which is first order in NO_2, and shows no dependence on NO_2F. The conclusion is that Mechanism 1 can NOT be correct; it is in conflict with experiment.

We next consider Mechanism 2, using the same general approach:

1. Write the rate law for the rds, which in this case is the first step in the mechanism. Rate = $k_1[NO_2][F_2]$ Further work is unnecessary here, because the rds rate law is already expressed in terms of concentrations of participants in the overall reaction. The implied rate law for Mechanism
2. Has the same form as the experimental rate law, and is therefore not in conflict with experiment.

Can we conclude, then, that Mechanism 2 is the correct mechanism for reaction 15–4–1? The answer is an emphatic NO. If a mechanism does not agree with experiment, it must be wrong, but just because a mechanism agrees with experiment does not mean that it is correct. A mechanism is a theory—it can never be proved. To underscore this point, consider the following possible mechanism for 15–4–1:

Mechanism 3

$$NO_2 + F_2 \rightarrow NO_2F_2 \ [k_1 \text{ slow}]$$

$$NO_2F_2 + NO_2 \rightarrow 2NO_2F \ [k_2 \text{ fast}]$$

This also is consistent with experiment, giving the implied rate law

$$\text{Rate} = k[NO_2][F_2]$$

based on the first step, which is the slow one. You might like to assess the validity of yet a fourth possible mechanism, given below:

Mechanism 4

$$1/2\ F_2 \rightarrow F \ [\text{fast}]$$

$$F + NO_2 \rightarrow NO_2F \ [\text{slow}]$$

We can now appreciate the significance of a statement made earlier: the experimental rate law tells us what atoms are contained in the reactants of the rate-determining step. The experimental rate law in Example 15–11, Rate = k $[NO_2][F_2]$, reveals that the reactants in the slow step must contain a total of 1 N atom, 2 O atoms, and 2 F atoms. Based on this, we can immediately rule out mechanism 1, for which the rds involves only one F atom; and mechanism 4, for which the same is true. Both mechanisms 2 and 3 have rds's containing the correct numbers of atoms, so neither is in conflict with experiment.

In general, chemists try to propose mechanisms that agree with the experimental rate law and with any other experimental evidence; and that are chemically reasonable. It is the assessment of chemical reasonability that requires experience. We consider one additional example of mechanism assessment.

Example. Several mechanisms might be considered for the overall reaction

$$2NO + O_2 \rightarrow 2NO_2$$

with rate law, Rate = $-d[O_2]/dt = k_{obs}[NO]^2[O_2]$. The symbol, k_{obs}, where "obs" stands for "observed", is commonly used for an experimentally measured rate constant.

Mechanism 1:

$$NO + NO \Leftrightarrow N_2O_2 \; [K_1 \text{ fast}]$$

$$N_2O_2 + O_2 \rightarrow 2NO_2 \; [k_2 \text{ slow}]$$

Mechanism 2:

$$NO + O_2 \Leftrightarrow OONO \; [K_1' \text{ fast}]$$

$$NO + OONO \rightarrow 2NO_2 \; [k_2 \text{ slow}]$$

Mechanism 3:

$$NO + O_2 \rightarrow OONO \; [k_1 \text{ slow}]$$

$$NO + OONO \rightarrow 2NO_2 \; [k_2 \text{ fast}]$$

Can any of these mechanisms be ruled out?

Solution: Mechanism 3 can be eliminated right away, because the reactants in the rds do not contain the correct numbers of atoms. The experimental rate law demands that there be two N atoms and four O atoms in the rds reactants; however, only one N and three O atoms are involved in the rds of mechanism 3. Mechanisms 1 and 2 show the correct atoms in the rds. You should confirm that the implied rate laws are

$$\text{Mechanism 1: Rate} = k_2K_1[NO]^2[O_2]$$

$$\text{Mechanism 2: Rate} = k_2K_1'[NO]^2[O_2]$$

both of which are consistent with experiment. Attempts to draw

Lewis structures for the intermediates, N_2O_2 and OONO, are shown in Figure. The intermediate OONO is a free radical (it has an odd electron), and should therefore react very rapidly with the free radical NO. It is thus unlikely that the second step of Mechanism 2 would be slow. Mechanism 1 is preferable, based on chemical reasonability. There is yet a fourth mechanism, involving a one-step termolecular process. It is currently believed that this is the correct one.

O=N—N=O   •O—O—N=O:

Let's summarize what we have just learned about reaction mechanisms:

- The steps in a mechanism are numbered from 1 to n. The subscripts on the elementary process rate constants are given the same numbers, with k_n being the forward rate constant and k_{-n} the reverse rate constant for step n. The steps must add to give the overall reaction.
- The mechanism often (but not always) involves at least one slow step. This is the rate-determining step.
- Any fast step preceding the slow step is treated as reversible, and assumed to come to equilibrium.
- To obtain the rate law, we
  - Write the rate law for the elementary rds;
  - Eliminate concentrations of intermediates using rapid pre–equilibrium expressions. This is necessary because the concentrations of intermediates are not usually observable.
- That the reaction proceeds in several steps having different rate constants can cause its rate law to deviate from the overall stoichiometry. However, the rate law tells us which atoms are involved in the reactants of the rate determining step, so is strongly suggestive of mechanism.
- Not all elementary step rate constants need appear in the implied rate law for the mechanism.
- Mechanisms can be disproven, but not proven.

TEMPERATURE DEPENDENCE OF RATE—KINETIC MOLECULAR THEORY

The rates of chemical reactions increase with increasing temperature, often dramatically. Coal reacts with oxygen at a negligible rate at room temperature, but burns spontaneously with the liberation of heat and light at 500 °C. A sirloin steak "cooks" (oxidizes) very slowly at room temperature, but requires only 20 minutes at 180°C. Reactions are accelerated by temperature increase because the elementary processes by which they occur take place more rapidly.

We begin this section with two questions:

1. Why are some elementary processes fast and others slow at a particular temperature?
2. Why are reaction rates (rate constants) so dramatically temperature dependent?

We can arrive at answers to these questions by looking in more detail at the collision process.

REQUIREMENTS FOR COLLISION TO LEAD TO REACTION

As discussed earlier, rate is proportional to the number of molecular collisions occuring per unit volume per unit time. For the decomposition of hydrogen iodide to hydrogen and iodine, shown in 15–5–1, assuming that [HI] = 0.001 M and a temperature of 500 °C, the rate would

$$2HI(g) \rightarrow H_2(g) + I_2(g)$$

be 5.8×10^4 M/s if all collisions led to formation of products. The observed rate, however, is only $1.2 * 10^{-3}$ M/s, $5 * 10^7$ times slower than it would be if all collisions between HI molecules were effective. Our conclusion must be that only a small fraction of molecular collisions lead to formation of products; *i.e.*, are*effective*. In general, for the elementary process,

$$A + B \rightarrow C + D$$

Rate = Number of effective collisions/volume-time, which should be proportional to the total number of collisions per unit volume per unit time. We call this total number of collisions Z. We can relate Z to the concentrations of A and B via equation 15–5–3:

$$Z = Z_0[A][B]$$

where Z_0 is the number of collisions per volume per time when A and B have 1 molar concentrations. The number of effective collisions is less than the total number of collisions for two reasons. *First, two colliding molecules must be properly oriented if they are to react.* Reaction requires that bonds within reactants be broken and that new bonds form between atoms in the colliding molecules. A bond cannot form between atoms in separate molecules unless they are near each other during collision. Figure shows two of the many possible collision orientations of CO and NO_2, which are reactants.

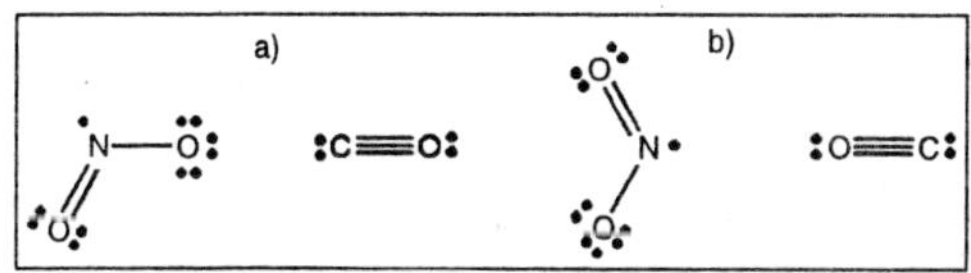

Fig. 10.8

$$CO + NO_2 \rightarrow CO_2 + NO$$

This bimolecular elementary process involves transfer of an oxygen atom from NO_2 to CO. For this to occur, it is necessary that an oxygen atom of NO_2 approach the carbon atom of CO during collision, so that a bond may form between them. This requirement is satisfied in the orientation in Figure. The relative orientation in brings the nitrogen atom of NO_2 and the oxygen atom of CO together, so cannot be effective in advancing the process. Similarly, many other possible orientations of these two molecules are unsuitable for transfer of an oxygen atom from N to C, so will not be effective. We symbolize the fraction of collisions having the proper orientation of molecules as p, and recognize that p is a small number.

Second, for reaction to occur, it is necessary that colliding molecules have some minimum total kinetic energy. This is called the activation energy, symbolize E_a. In almost all cases, successful reaction requires that existing bonds be broken and new bonds made. Two bonded atoms lie at a minimum in a potential energy curve of the type discussed and pictured in Figure.

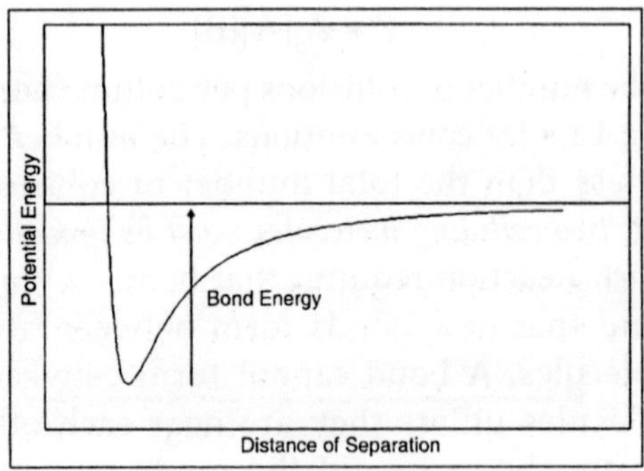

Fig. 10-9 Potential Well

For the bond to break, an input of energy at least equal to the depth of the well is required. This energy is drawn from the total kinetic energy possessed by the colliding species. If the kinetic energy is insufficient to overcome the bonding interactions, the reactant molecules separate without change. Even if the total kinetic energy is sufficient for reaction, the collision orientation may still be inappropriate for reaction.

Both factors—sufficient kinetic energy and correct orientation—are necessary for reaction to occur. Assuming that the minimum energy is available and the orientation correct, bonds will be ruptured, new bonds will form with the release of energy, and new molecules will separate from the collision center. Between reactants and products, both of which are relatively low in potential energy, lies a state of maximum potential energy called the *transition state*.

The arrangement of atoms that exists in this state is called the *activated complex*. We may think of it as a "molecule" in which chemical bonds are in flux, some being broken and others being made. We can depict this situation graphically in terms of a *reaction coordinate diagram*, which is a plot of potential energy versus the progress of reaction, which is usually called the "reaction coordinate." A reaction coordinate diagram for the bimolecular elementary process shown in Figure.

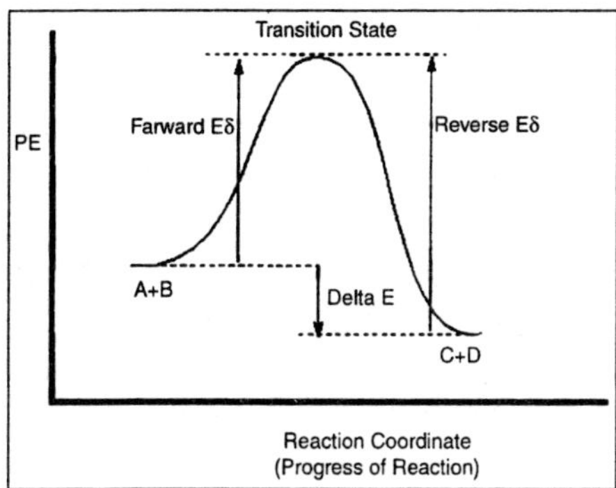

Fig. 10-10

Potential energy is plotted on the vertical axis, with the reaction coordinate on the horizontal axis. The reactants, A and B are located to the left on the reaction coordinate. As reaction proceeds, potential energy rises to the maximum at the transition state, then falls again as bond formation in the product species is completed and the product molecules separate.

The activation energy, E_a, for conversion of A and B to C and D (the forward reaction) is the difference in potential energy between the transition state and the reactants. This is labelled in the diagram. In drawing this diagram for the generic process 15–5–2, it has been assumed that products have lower potential energy than reactants; that is, that the process is exothermic. Thus DE for the process is negative. The activation energy for the reverse process, also indicated in the diagram, is the difference in PE between the transition state and the potential energy of C + D. It is clear from the figure that ΔE for the reaction is the difference between the activation energies for the forward and reverse processes. This is generally true.

$$\text{(15–5–5): } E_a(\text{forward}) = \text{PE(activated complex)} - \text{PE(A+B)}$$

$$E_a(\text{reverse}) = \text{PE(activated complex)} - \text{PE(C+D)}$$

$$E_a(\text{forward}) - E_a(\text{reverse}) = \Delta E \text{ for elementary process}$$

Every elementary process can be depicted as a single-humped potential energy curve, with an activated complex at the maximum. Letting the fraction of collisions with sufficient kinetic energy to react be f, then

(15–5–6): Rate = Number of effective collisions/volume-time = pfZ

We are now in a position to answer the first of the two questions posed at the beginning of this section. Some elementary processes are slower than others either because they have very restrictive orientation requirements, corresponding to a very small value for p; or they have high potential energy barriers (E_a), corresponding to a small value for f. Next we consider the effect of an increase in temperature. The distribution of kinetic energies in a collection of molecules is governed by the Maxwell-Boltzmann distribution, first introduced.

The M-B distribution curve for temperature T_1 is shown in Figure. The value of the activation energy for elementary process has been indicated on the graph. The fraction of molecules with kinetic energy greater than E_a is given by the area under the M-B curve to the right of E_a. As temperature is increased to some new value T_2, the M–B curve shifts to the right and flattens, indicating the increase in both the average molecular kinetic energy and the breadth of the distribution. The high temperature curve is also shown in the figure. Notice that the area under the curve to the right of E_a is much larger for the high-temperature curve, leading to a dramatic increase in the number of effective molecular collisions. Quantitatively, the fraction of effective collision can be expressed.

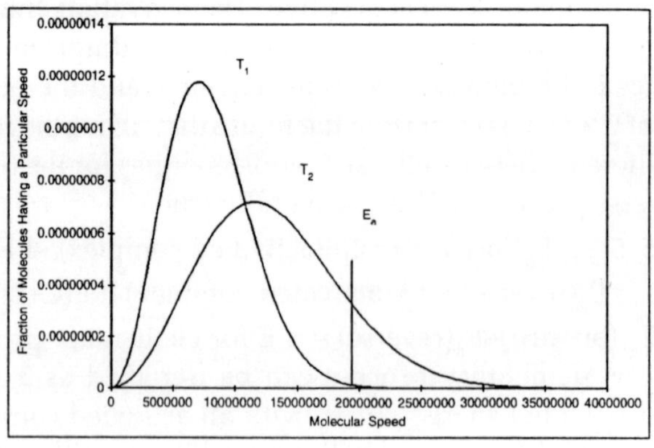

Fig. 10-12 Maxwell-Boltzmann Distribution

$$f = \text{Area under curve right of } E_a \ \alpha \ e^{-Ea/RT}$$

Thus,

$$\text{Rate} = pfZ = pe^{-Ea/RT}Z_o[A][B]$$

Equation 15-5-8 has the form of a rate law if we identify the terms preceding the concentrations of A and B as the rate constant, k:

$$k = pZ_oe^{-Ea/RT} = Ae{-}^{Ea/RT}$$

where A contains information about the orientation requirements. Taking the logarithm of both sides of 15-5-9 gives 15-5-10, called the Arrhenius equation.

$$\ln k = \ln A - E_a/RT$$

The value of the activation energy for a process is obtained by measuring the rate constant at each of several temperatures, and constructing a plot of lnk versus 1/T. The plot should be linear, with slope = $-E_a/R$ and intercept = lnA. We are now able to offer an answer to the second of the two questions posed at the outset: reaction rates are temperature dependent because of the requirement of a minimum kinetic energy for effective collisions.

Example. Aqueous hydrogen peroxide and iodide ion react according to the following equation.

$$H_2O_2(aq) + 3I{-}(aq) + 2H^+(aq) \rightarrow I_3{-}(aq) + 2H_2O$$

The reaction is first order in peroxide and first order in iodide, second order overall. The reaction was studied at several temperatures, with the following results. What is the activation energy for the reaction?

T, °C	k_{obs}, $M^{-1}s^{-1}$
2	3.03 * 10^{-3}
17	1.0 * 10^{-2}
50	7.6 * 10^{-2}

Solution. The activation energy can be obtained from the slope of a plot of ln k_{obs} versus reciprocal Kelvin temperature. The plot is in Figure. The activation energy obtained from the slope is 53.6 kJ/mole.

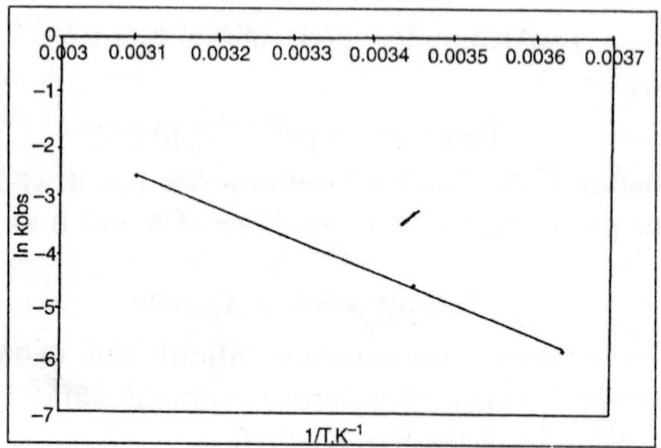

Fig. 10-13 Arrhenius Plot For Example 15-10

OVERALL (MULTISTEP) REACTIONS

Thus far we have discussed the temperature dependence of a single elementary process. Similar considerations apply to overall (net) reactions consisting of several steps.

We make several general statements about reaction coordinate diagrams for net reactions before looking at a specific example:

- The number of maxima in a reaction coordinate diagram is equal to the number of steps in the mechanism.
- The slowest step usually has the highest E_a.
- The rate law for the overall reaction indicates the atoms present in the activated complex for the slowest elementary step.
- Intermediates occur at minima in the reaction coordinate diagram.
- The E_a measured experimentally for the multistep reaction is the difference between the potential energy at the highest point in the diagram and the potential energy of the reactants.

Example: For the reaction

$$2NO_2 + F_2 \rightarrow 2NO_2F$$

the rate law is $-d[F_2]/dt = k_{obs}[NO_2][F_2]$. Propose a mechanism for the reaction, and draw an appropriate reaction coordinate diagram.

Solution: The rate law tells us that the activated complex of the slow step contains two oxygen atoms, two fluorine atoms, and one nitrogen atom. This suggests the following mechanism:

$$NO_2 + F_2 \rightarrow NO_2F + F \; [k_1 \text{ slow}]$$

$$F + NO_2 \rightarrow NO_2F \; [k_2 \text{ fast}]$$

Because reaction involves the breaking of the F-F bond and formation of a bond between N and F, the structure shown in Figurea is a reasonable arrangement of atoms in the activated complex. Finally, Figure shows the reaction coordinate diagram corresponding to the proposed mechanism. There are two "humps" (maxima), one corresponding to each elementary step in the mechanism; the intermediate, F, is located at the minimum between maxima; the activation energy for the overall reaction is indicated; and the overall reaction is exothermic. The activation energy for the first step is substantially larger than that for the second step, consistent with the first step being rate-determining. Thus the reaction coordinate diagram reveals by its profile which is the slow step in the overall reaction.

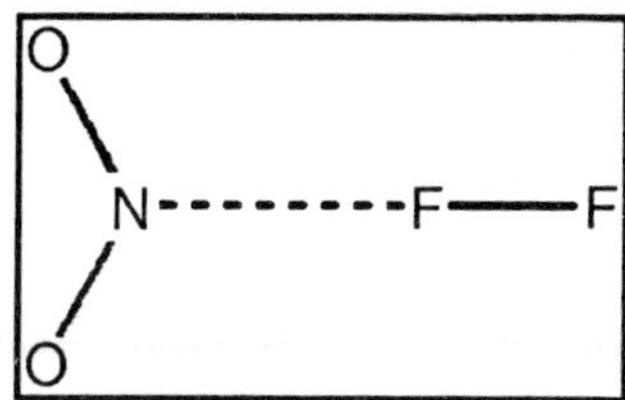

Fig. 10.15

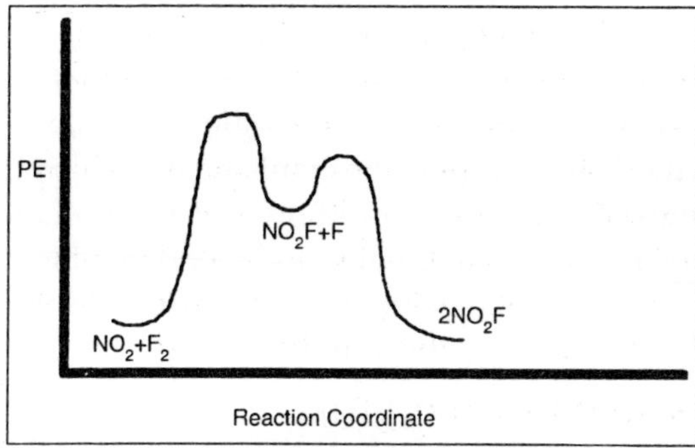

Fig. 10.16

Example: The reaction coordinate diagram for an unspecified reaction is shown in Figure. From the details of the diagram, say as much as possible about the net reaction to which it corresponds.

Solution: Because there are three maxima in the diagram, the mechanism must consist of three steps. Step 2 must be the rds, because it has the largest E_a. There is one activated complex for each step, for a total of three. There are two minima, indicating two intermediates. The experimental rate law for the reaction should indicate the composition of the activated complex of the second step. The overall reaction is exothermic, because products are lower on the energy axis than reactants. Finally, the measured E_a is the energy difference between the highest maximum in the diagram and the reactants. It is shown in the diagram.

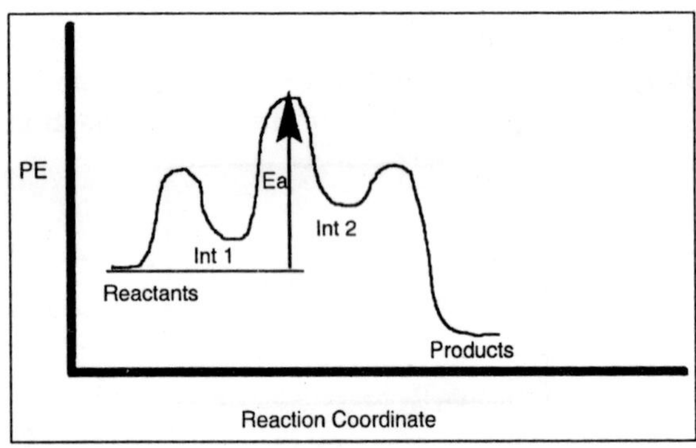

Fig. 10.17

The temperature dependence of the observed rate law for an overall reaction consisting of several steps is usually qualitatively similar to that for a single elementary process. However, there are complex mechanisms for which this is not the case. Indeed, examples are known of reactions for which the observed rate constant is either independent of temperature or decreases with increasing temperature. We will not be concerned with such situations in this book.

KINETICS AND EQUILIBRIUM

The connection between the equilibrium constant for a

reaction and the rate constants of the elementary steps by which it occurs is an important one. We begin by stating the relationship. We will then demonstrate, for a specific case, that it is true.

The equilibrium constant for a net reaction is the ratio of the product of the forward rate constants for all steps in the mechanism to the product of the reverse rate constants for all steps in the mechanism:

$$K_{eq} = k_1k_2k_3.../k_{-1}k_{-2}k_{-3}...$$

We now show for a specific net reaction and a specific (proposed) mechanism that this statement is true. Consider again the reaction of NO_2 and F_2 to produce NO_2F, reaction 15–4–1:

$$2NO_2 + F_2 \Leftrightarrow 2NO_2F$$

We have written the reaction with the double arrow because we are now interested in the forward and reverse rates at equilibrium. At equilibrium, the overall forward rate and overall reverse rate must be equal:

$$Rate_f = Rate_r$$

This must be true regardless of the detailed pathway by which the reaction occurs. All available experimental evidence indicates that the mechanism for 15–4–1 is the two-step process in 15–4–2a and b:

$$NO_2 + F_2 \Leftrightarrow NO_2F + F\ [k_1, k_{-1}]$$

$$NO_2 + F \rightarrow NO_2F\ [k_2, k_{-2}]$$

When the overall reaction is in equilibrium, so must each elementary step be. Consequently, we have used double arrows in the elementary steps as well. The equality of the forward and reverse rates of the individual steps is expressed mathematically as follows.

$$k_1[NO_2]_e[F_2]_e = k_{-1}[NO_2F]_e[F]_e$$

$$k_2[NO_2]_e[F]_e = k_{-2}[NO_2F]_e$$

Eliminating the concentration of the intermediate, F, gives.

$$[NO_2F]_e^2/[NO_2]_e[F_2]_e = k_1k_2/k_{-2}k_{-2} = K_{eq}$$

The statement made at the outset is thus seen to be true for this specific case: *The equilibrium constant is the ratio of the product of forward rate constants to the product of reverse rate constants.* For the general overall reaction occuring by an n-step mechanism,

$$K_{eq} = k_1k_2...k_n/k_{-1}k_{-2}...k_{-n}$$

Objection is frequently made to the method of derivation that we have just used, because the mechanism for is not known with certainty; it is and will remain a hypothesis, however well based in experiment. If it is found that the two step mechanism in 15–4–2 is indeed NOT correct, then the relationship in is invalid. This is certainly true. However, even if the mechanism above is incorrect, must occur by some mechanism.At equilibrium each step in the mechanism must be balanced in rate, enabling us to eliminate concentrations of any and all intermediates just as we did above, to arrive at a modified version of 15–6–2. The validity of the general relationship in 15–6–3 is independent of the details of mechanism.

Thus the intuitively-expected connection between k and K does indeed exist. In fact, there are many examples in which the equilibrium constant for a reaction has been obtained from kinetics studies. This process can never be reversed however; it is not possible to obtain k by performing equilibrium studies. Before leaving this matter, we make one more very important point. There is a tendency to believe that reactions with large equilibrium constants are fast, whereas those with small equilibrium constants are slow. However, neither of these beliefs is true. A large value for K_{eq} means that the product of forward rate constants is much larger than the product of reverse rate constants. It does not follow, however, that the forward rate constants are Large. It means only that they are larger than the reverse rate constants. $k_1k_2...k_n >> k_{-1}k_{-2}...k_{-n}$ does not mean that $k_1, k_2..., k_n$ are large in the absolute sense.

A large K_{eq} can result from the ratio of a slow forward rate to an even slower reverse rate.

CATALYSIS

A number of chemical reactions that ordinarily occur slowly can be induced to occur more rapidly by the addition of a suitable substance called a *catalyst*. We will define a catalyst as a substance that speeds up a reaction without itself being consumed or chemically changed in the overall reaction process. This definition is in practice rather restrictive, because eventually, all real catalysts become deactivated by "poisoning" or via irreversible structural changes; thus they cannot function indefinitely. For our purposes, here, however, this definition will serve. The reactant molecule that is affected by the catalyst is called the *substrate*. The process by which a catalyst effects its action is called *catalysis*. Catalysis is one of the most intensely studied areas in science because it is of tremendous biological and industrial importance. We begin our exploration of catalysts by discussing how they work.

THE MODE OF CATALYST FUNCTION

A catalyst functions either by entering into a slow step of the uncatalyzed reaction mechanism, or by creating an entirely new mechanism for the reaction. In either case, it is thought that the catalyzed pathway has a lower activation energy than the uncatalyzed path. We can illustrate catalytic action generically in terms of the following reaction scheme, where A, B, D, and F are reactants and products in the overall reaction, and C is a catalyst for the reaction.

$$A + B \rightarrow D + F \text{ [overall]}$$

Uncatalyzed mechanism:

$$A + B \rightarrow AB \text{ [slow]}$$

$$(15\text{–}7\text{–}2b): AB \rightarrow D + F$$

Catalyzed mechanism:

$$A + C \rightarrow CA \text{ [fast]}$$

$$CA + B \rightarrow CAB$$

[less slow than first step of uncatalyzed mechanism]

$$CAB \rightarrow D + CF \text{ [fast]}$$

$$CF \rightarrow C + F \text{ [fast]}$$

The effect of the catalyst on the activation energy of the slow step of the reaction is shown schematically in Figure. As shown in the figure, the catalyst affects only E_a; it does not affect the energy of either the reactants or products of the reaction. The intermediate, CAB, in which both reactants are bound to the catalyst, is of particular interest. A possible (and fairly common) structural motif for this intermediate is shown in Figure. Notice that, although A and B are bound to different sites on the catalyst, C, they are in proximity and can effectively interact. Binding of A and B to C causes shifts in electron density that may facilitate bond breaking within A and/or B and bond formation between A and B or fragments of them. Thus the catalyst provides an organizing center for A and B, facilitating their interaction. In this manner the catalyst can overcome the orientation factor discussed earlier.

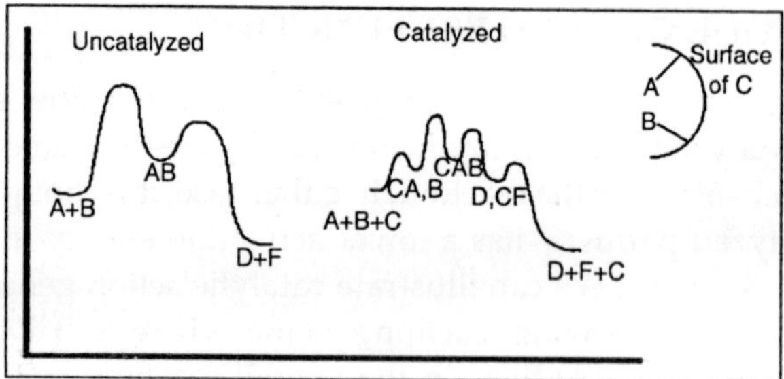

Fig. 10-18

As soon as the molecule of catalyst is regenerated in the last step of the 4-step mechanism, it may bind another molecule of A and proceed once again through the sequence, converting A and B to D and F. This sequence is repeated a large number of times, so that generally only a small amount of catalyst is required to convert a substantial quantity of reactant to product. Chemists call the process by which the catalyst cycles through the reaction over and over again a catalytic loop, and represent it as in Figure. The unbound form of the catalyst, C, is placed at the 12 oclock position of the loop, and the various forms in which the catalyst is found are placed more or less evenly around the remainder of the loop. Arrows show the direction of reaction around the loop, with reactants brought in from outside the loop, and products ejected out of the loop. This is a very effective visual presentation of catalyst action. The number of times that a molecule of catalyst cycles through the loop per unit time is called the *turnover number* of the catalyst. An effective measure of turnover number is moles product produced per mole catalyst present per time.

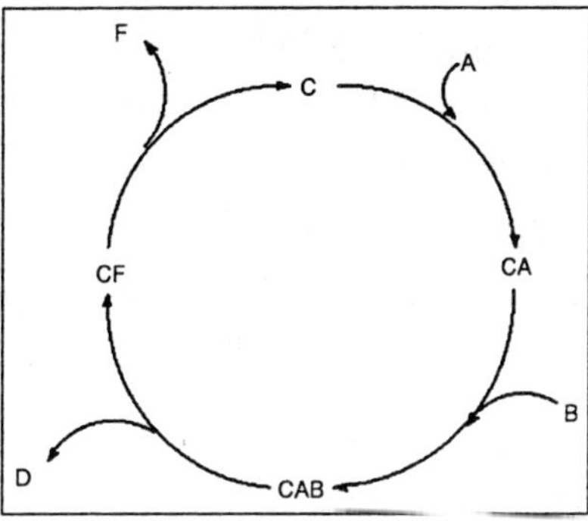

Fig. 10.19

Let's look at a specific example of the effect of a catalyst on a simple reaction, from which we can draw some general conclusions.

Example: The reaction of ethanol (ethyl alcohol) with bromide ion to produce ethyl bromide and hydroxide ion is shown in 15–7–4.

$$C_2H_5OH + Br- \rightarrow C_2H_5Br + OH^-$$

The reaction is quite slow in neutral or basic solution, but occurs quite rapidly in acidic solution because it is catalyzed by the hydronium ion, H_3O^+. Discuss the mechanisms for the uncatalyzed and catalyzed processes.

Solution. The uncatalyzed reaction is thought to occur in a single elementary step, in which bromide ion attacks the hydroxyl carbon atom of ethanol while the hydroxide group simultaneously departs:

$$Br^- + C_2H_5OH \rightarrow \text{activated complex} \rightarrow C_2H_5Br + OH^-$$

$$\text{Rate} = k_{obs}[C_2H_5OH][Br^-]$$

This mechanism is proposed based on the experimental rate law, which is overall second order. The observed rate law for the catalyzed process is

$$\text{Rate} = k_{obs}[Br^-][C_2H_5OH][H_3O^+]$$

The following three-step mechanism is consistent with this rate law if k_{obs} is identified with k_2K_1:

$$(15\text{–}7\text{–}6a): C_2H_5OH + H_3O^+ \rightarrow C_2H_5OH_2^+ + H_2O \text{ [fast, } K_1]$$

$$(15\text{–}7\text{–}6b): C_2H_5OH_2^+ + Br^- \rightarrow C_2H_5Br + H_2O \text{ [slow } k_2]$$

$$(15\text{–}7\text{–}6c): 2H_2O \rightarrow H_3O^+ + OH^-$$

The reaction coordinate diagrams for the uncatalyzed and catalyzed pathways are shown in Figure.

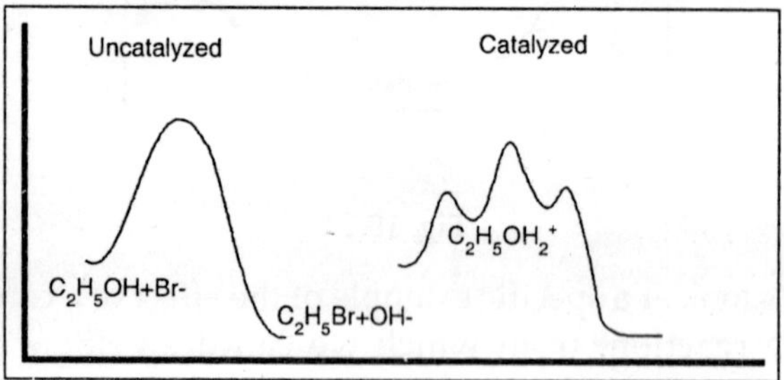

Fig. 10.20

Several general statements about catalysis are evident from this example:

- Because the catalyst does not appear in the overall equation for the reaction, it does not affect the equilibrium; it affects only the kinetics. The catalyst speeds up not only the forward reaction, but also the reverse, by the same factor. The position of equilibrium remains the same.
- The activation barrier is lower in the catalyzed mechanism. Even though the catalyzed path involves more steps, the overall reaction occurs more rapidly.
- The concentration of the catalyst appears in the rate law, even though the catalyst does not appear as a reactant or product in the overall equation.

- Although a catalyst and an intermediate share the property of not appearing in the overall equation, they are different types of things. First, the catalyst appears in the rate law; an intermediate cannot. Second, the catalyst appears first as a reactant, then is generated later as a product. An intermediate appears first as a product, then later as a reactant.

EXAMPLES OF CATALYSIS

We will now briefly discuss several examples of catalysis that are actually used on a mammoth scale in the chemical industry. Each of these processes, in its own way, has a major impact on the quality of our lives.

The Polymerization of Ethylene

Ethylene is a very simple molecule with formula C_2H_4. It is a gas, obtained as a byproduct during the catalytic cracking of petroleum. Under certain conditions, ethylene molecules can be made to join together end-to-end to form very long chain-like molecules of *polyethylene*, so called because it consists of m

$$(15-7-7): 2bCH_2 = CH_2 \xrightarrow{Ticl_3} -(CH_2\,CH_2\,CH_2\,CH_2)_{n}-$$

Even at high temperature and pressure of ethylene, this process occurs negligibly slowly. In the presence of a small amount of a modified form of $TiCl_3$ (called tickle-3 in the plastics industry), however, it occurs rapidly at only moderately high temperature and pressure. The process is referred to as Ziegler-Natta catalysis after its two coinventors, who jointly received the Nobel Prize in Chemistry in 1963. Since the discovery of this process in the 1950's, the entire plastics industry has developed and grown to huge proportions. To this day, the mechanism of Ziegler-Natta catalysis is not fully understood. Mechanistic studies are difficult for several reasons, one of which is that the process is *heterogeneous*; that is, the catalyst ($TiCl_3$) and substrate (ethylene) are in different phases. The catalytic process takes place on the surface of crystals of $TiCl_3$, which rapidly become covered with and blocked from view by the resulting polyethylene. Much effort is ongoing in the US chemical industry to develop more efficient and easily handled Ziegler-Natta catalysts.

The Production of Sulfuric Acid

Year after year, sulfuric acid ranks first on the list of the top ten chemical substances produced in the United States: billions of pounds are produced annually. Sulfuric acid is synthesized by the so-called *Contact Process*, which involves the four sequential steps below:

$$S + O_2 \rightarrow SO_2 \text{ [fast]}$$

$SO_2 + O_2 \rightarrow SO_3$ [slow, because it occurs by a termolecular elementary process catalyzed by V_2O_5]

$$SO_3 + H_2SO_4 \rightarrow H_2S_2O_7 \text{ [fast]}$$

$$H_2S_2O_7 + H_2O \rightarrow 2H_2SO_4 \text{ [fast]}$$

The reaction of SO_3 with water is very exothermic and causes extensive spattering and production of a fine mist of highly acidic water. For this reason, direct reaction of SO_3 with water in the third step is impractical. Instead, SO_3 is bubbled into pure sulfuric acid, with which it reacts smoothly to give fuming sulfuric acid, $H_2S_2O_7$.

This can then be treated with the stoichiometrically correct amount of water to give sulfuric acid. The contact process has been so perfected that sulfuric acid is very inexpensive to produce. Consequently, it is used in any industrial process requiring acid. It finds it major uses in the production of phosphate fertilizers; in paper manufacture; in the petroleum industry; in steel production; and in the production of detergents. The importance of these products in our lives is obvious.

Catalytic Converters

For some years now, catalysts have been placed within the exhaust systems of automobiles to reduce the amount of poisonous or otherwise harmful emissions. These noble metal catalysts (based on platinum and palladium) carry out a dual function. First, they facilitate oxidation of carbon monoxide, resulting from incomplete hydrocarbon combustion, to carbon dioxide:

$$CO(g)+1/2O_2(g)\xrightarrow{Pt}CO_2(g)$$

Second, they catalyze decomposition of nitric oxide, produced during engine operation and oxidized rapidly to toxic NO_2 by atmospheric oxygen, to N_2 and O_2. Unfortunately, catalytic converters also facilitate oxidation of SO_2 to SO_3, which is the precursor of acid rain. Low-sulfur petroleum distillates are therefore essential. The Haber Process. Vast quantities of ammonia are synthesized each year for use as fertilizer. Currently the most efficient process for ammonia synthesis is the Haber-Bosch Process, developed during the ten-year period preceding 1913, in which nitrogen and hydrogen react directly at high temperature and pressure and in the presence of an activated iron catalyst to form ammonia.

$$N_2(g)+3H_2(g)\xrightarrow{Fe}2NH_3(g)$$

A catalyst and high temperature are necessary to cause the reaction to go at a reasonable rate. Unfortunately, high temperature makes the exothermic reaction less favoured, so very high pressure is used to favour products.

Even with conditions optimized, reaction is incomplete, and unreacted hydrogen gas is recycled for maximum efficiency of ammonia production. Nitrogen fertilizers produced from ammonia are largely responsible for the incredible growth in agricultural production over the decades since the First World War, when the process was first put on line in Germany. Ironically, the original motivation for development of the process was the requirement of explosives for the war effort.

The processes discussed above have at least three features in common. First, catalysis is heterogeneous; the catalyst is in all cases a solid, interacting with the substrate in the gas phase. Second, the mechanisms for these processes are incompletely understood. Thus catalysts are used successfully on a huge scale, even though we do not understand how they work. Third, in all cases the catalyst involves a transition metal, from the d block of the periodic table. Transition metals are often versatile catalysts because they are flexible in coordination number (that is, the number of atoms, ions, or molecules to which they may bind in a Lewis acid-base interaction); in stereochemistry (that is, in the shapes that their adducts assume); and in oxidation state (that is, in the charge that they carry). Nature has chosen transition metals to serve as the centerpieces for many of its catalysts, the enzymes, most probably for these same reasons.

COMPLEX REACTIONS

Parallel Processes, Consecutive Processes, and Reversible Processes. Thus far we have discussed relatively simple reactions, which occur by mechanisms that involve one step that is substantially slower than the others, and which go to completion. In this situation, rate laws may be readily written from the mechanism and compared with the laboratory-determined rate law.

As you might expect, most reactions are not so simple. In this section, we briefly describe three frequently encountered complications: parallel reactions (two or more simultaneous processes leading to the same or different products); consecutive reactions having comparable rates; reversible reactions (those in which the reverse process plays an important role in determining the observed rate).

PARALLEL REACTIONS

Parallel reaction paths may be represented schematically as in 15–8–1.

$$A + B \rightarrow P\ [k_1]$$
$$A + C \rightarrow Q\ [k_2]$$

To simplify discussion, we assume that each reaction occurs by a single bimolecular elementary step. If so, then we may write the rate laws directly. The rate of disappearance of A is the sum of the rates at which it is used in 15–8–1a and 15–8–1b:

$$\text{Rate} = -d[A]/dt = k_1[A][B] + k_2[A][C] = \{k_1[B] + k_2[C]\}[A]$$

If pseudo-order conditions are imposed, this simplifies to the pseudo first-order rate law in 15-8-3.

$$\text{Rate} = -d[A]/dt = (k_1' + k_2')[A]$$

where $k_1' = k_1[B]$ and $k_2' = k_2[C]$. A plot of Rate versus [A] is linear, with slope $k_1' + k_2'$. Thus if one follows the reaction rate by monitoring [A], only the sum of the two rate constants is obtained. If it is possible to monitor [P] or [Q], k_1' or k_2' may be separately determined, and the other one calculated by difference.

Example. The overall reaction below is thought to occur by two different pathways operating simultaneously.

$$Pt(NH_3)_2Cl_2 + H_2O \rightarrow Pt(NH_3)_2Cl(H_2O)^+ + Cl^-$$

The first pathway involves attack by H_2O at the platinum atom, with simultaneous loss of Cl–. Product is thus formed in a single elementary step with the same equation as the overall reaction, with rate constant k_1. In the second pathway, Cl^- leaves the platinum compound in a slow first step; this is followed by a rapid second step in which H_2O forms a bond with platinum:

$$Pt(NH_3)Cl_2 \rightarrow Pt(NH_3)Cl^+ + Cl^- \; [k_1, \text{slow}]$$

$$Pt(NH_3)_2Cl^+ + H_2O \rightarrow Pt(NH_3)_2Cl(H_2O)^+ \; [k_2, \text{rapid}]$$

What rate law will be exhibited if this hypothesis is correct?

Solution. The overall rate law is the sum of the rate laws for the two parallel processes:

Rate = Rate mechanism 1 + Rate mechanism 2

$$= k_1[Pt(NH_3)_2Cl_2][H_2O] + k_2[Pt(NH_3)_2Cl_2]$$

(Be sure that you understand why the rate laws for the two mechanisms have these forms.)

CONSECUTIVE REACTIONS

We have previously examined systems consisting of two consecutive reactions under what might be called limiting conditions: either the first reaction was much more rapid than the second and could be treated as an equilibrium; or the first reaction was much slower than the second and completely determined the overall reaction rate. It is instructive to examine how the concentrations of A, B, and C vary with time in these situations, as well as the one in which the two steps have comparable rate constants.

Consider the following two step mechanism:

$$A \rightarrow B\ [k_1]$$
$$B \rightarrow C\ [k_2]$$

We present three situations involving different relative sizes of k_1 and k_2. We assume for simplicity that only A is present initially; [B] = [C] = 0 at t = 0. When $k_1 >> k_2$, reaction goes to completion before 15-8-4b gets under way. When $k_1 = 20k_2$, the concentrations of A, B, and C vary as shown in Figure. A rapidly disappears and is replaced by B; B then very slowly converts to C.

If the reaction rate is monitored by measuring the time dependence of [C], only the rate constant k_2 is obtained. On the other hand, if [A] is monitored, its rapid disappearance is governed by k_1, which is obtainable from either a plot of ln [A] versus time, or a plot of -d[A]/dt vs [A]. When $k_1 << k_2$, B is formed very slowly via 15–8–4a, but is then immediately consumed by 15–8–4b.

The concentration of B thus never rises much above zero, and C appears at the same rate that A disappears. This corresponds to the second set of limiting conditions mentioned above. When $k_2 = 20k_1$, the concentrations of the 3 species vary as shown in Figure. The rate constant determined experimentally is k_1, the smaller of the two. In this situation, it is very difficult to measure k_2 as [B] is never appreciable.

Only if B can be prepared independently under conditions where it does not convert to C can the rate constant, k_2, be measured.

There are many situations in which k_1 and k_2 are approximately equal. We illustrate these by looking at the case in which $k_1 = k_2$. The time dependence of the concentrations of A, B, and C is presented in Figure. At the very beginning of reaction, t = 0, the rate of the second reaction is zero because [B] = 0. The initial rate is the rate of disappearance of A, from which the rate constant k_1 can be determined.

Before reaction has progressed very far, however, [B] has risen sufficiently that the second process occurs simultaneously with the first. This tends to drain the concentration of B as it is converted to C. In the Figure, we see [B] rise to a maximum, then fall off because its rate of production from A slows as A is consumed. At any time, it must be true that [A] + [B] + [C] = constant = $[A]_o$. If [C] is monitored and [A] is calculated from the value of k_1, $[A]_o$, and t, [B] can be calculated as a function of time. It is then possible to determine k_2 from appropriate plots.

Finally, the mechanism in 15-8-4a and b can be modified slightly to allow reversibility in the first (but not the second) step:

$$A \Leftrightarrow B\ [k_1, k_{-1}]$$
$$B \rightarrow C\ [k_2]$$

(When $k_{-1} = 0$ this mechanism reduces to the one in 15–8–4a and b). Figure 10.20 shows the time dependence of the concentrations of A, B, and C assuming k_1, $k_{-1} >> k_2$. This corresponds to the first limiting situation mentioned at the beginning of this section. Figure shows the concentration behaviour for a short time early in the reaction. It is very clear from this figure that A and B rapidly come to equilibrium and maintain a constant concentration ratio thereafter. The total of their concentrations is then very slowly reduced as B converts to C. Figure shows the course of reaction over a longer time.

The plots in Figures 10.18 to 10.20 were obtained using finite difference methods to calculate the concentrations of the three species, A, B, and C, at many times during the course of reaction. These repetitive calculations were carried out using an electronic spreadsheet. Table–2 contains the data used in plotting outlines the method of calculation programmed into the spreadsheet. It would be instructive for you to set up your own spreadsheet according to the information in to see that you can produce plots similar to those in Figures 10.18 to 10–20.

REVERSIBLE REACTIONS

Our last example of complexity in chemical dynamics will focus on reversible reactions: those that approach an equilibrium position in which appreciable amounts of both reactants and products are present. The simplest such process is shown in 15/08–6.

$$A \Leftrightarrow B\ [k_1, k_{-1}]$$

k_1 and k_{-1} are the rate constants for the forward and reverse processes, respectively, which are both assumed to be first order. The rate law in 15–8–7 follows:

$$Rate = -d[A]/dt = k_1[A]-k_{-1}[B]$$

This equation states that the rate of change in the concentration of A is the difference between its rate of consumption via the forward process and its rate of production via the reverse process. If the initial concentration of B is 0, then at any point during the process, $[B] = [A]_o-[A]$.Equation 15-8-7 becomes

$$-d[A]/dt = (k_1 + k_{-1})[A]-k_{-1}[A]_o$$

This can be integrated, but the resulting expression is not readily plottable. Thus, even though the reaction is first-order, a plot of ln[A] versus t will not be linear. To eliminate the troublesome $k_{-1}[A]_o$ term, we recognize that 15–8–8 applies at all points during reaction, including equilibrium. Letting $[A]_e$ represent the equilibrium concentration of A, 15–8–9 holds when the reaction reaches equilibrium:

$$-d[A]_e/dt = (k_1 + k_{-1})[A]_e-k_{-1}[A]_o$$

Subtracting 15–8–9 from 15-8-8 eliminates the $k_{-1}[A]_o$ term, giving,

$$-d([A]-[A]_e)/dt = (k_1 + k_{-1})([A]-[A]_e) = k_{tot}([A]-[A]_e)$$

This equation is readily integrated in the variable $[A]-[A]_e$, which is the *deviation* of the concentration of A from the equilibrium value. Thus 15–8–10 says that in a reversible first-order process, the approach to equilibrium is governed by first-order kinetics. A plot of ln $([A]-[A]_e)$ against t will be linear, with slope $-k_{tot}$, the sum of the forward and reverse rate constants.

We make several observations about reversible first-order processes:

- The rate constant for approach to equilibrium is the sum of forward and reverse rate constants because equilibrium is approached by simultaneous reaction in both directions.

- When 15-8-6 is irreversible, $k_{-1} = 0$ and $[A]_e = 0$; thus 15-8-10 reduces to 15–3–6 (with n = 1).
- When $k_1 \gg k_{-1}$, the measured rate constant, k_{tot}, is essentially equal to k_1. The reverse process is unimportant until the reaction is nearly at equilibrium, because $k_{-1}[B]$ remains small. At equilibrium, the forward and reverse reactions, given by the following expressions, are equal:

$$\text{Rate}_f = k_1[A] = \text{Rate}_r = k_{-1}[B]$$

Both rates are small since at equilibrium [A] and k_{-1} are both small. Similar considerations hold when $k_1 \ll k_{-1}$. In this case, the measured rate constant is essentially k_{-1}.

Let's examine a particular example of a reversible first order process.

Example: The kinetics of the reaction below were monitored by following the change in absorbance with time of R.

$$R \Leftrightarrow P$$

The following data were obtained.

Time, Min	Absorbance Due to R
1	.789
2	.648
4	.452
6	.333
8	.261
10	.217
12	.191
60	.150

Solution. Determine the rate constant for approach to equilibrium. The absorbance value at 60 minutes represents the equilibrium value, since enough time has elapsed for the reaction to have reached equilibrium. Thus $A_e = 0.15$. A value of $A-A_e$ is calculated for each time and a plot of $\ln(A-A_e)$ constructed. Figureshows such a plot. The linearity of the plot substantiates a first order approach to equilibrium. The slope of the plot gives the rate constant, 0.27 min^{-1}. This is actually the sum of the forward and reverse rate constants for the reaction. More complicated processes can be made to conform to 15–8–10 by imposing pseudo-order conditions. For example, reaction 15–8–11 involves forward and reverse processes that are second order:

$$A + B \Leftrightarrow C + D\ [k_1, k_{-1}]$$

Suppose that initially $[B]_o$ and $[D]_o$ are both made much larger than $[A]_o$, and $[C]_o = 0$. Then the concentrations of B and D will remain constant during the reaction, and A will be converted to C in a pseudo-first-order process in which the forward rate constant, k_f, is equal to $k_1[B]$; and the reverse rate constant, k_r, is equal to $k_{-1}[D]$. By analogy with 15-8-10, we write the rate law for approach to equilibrium:

$$\text{Rate} = -d([A]-[A]_e)/dt = (k_f + k_r)([A]-[A]_e) = k_{tot}([A]\ [A]_e)$$

k_{tot} is studied first as a function of [B] with [D] held constant, in which case 15-8-13 applies:

$$k_{tot} = k_1[B] + k_r$$

k_1 can be obtained from a plot of k_{tot} versus [B]. Then k_{tot} is studied as a function of [D] with [B] held constant, when 15–8–14 applies:

$$k_{tot} = k_f + k_{-1}[D]$$

k_{-1} is obtained from a plot of k_{tot} versus [D].

ENZYMES AS CATALYSTS

Even the simplest organism is powered by thousands of chemical reactions, participating in a complex, interconnected web of chemical activity.

On their own, most of these reactions occur with rates so slow as to seem inconsistent with life. In the organism, however, they take place briskly and *specifically* (*i.e.*, only the desired products are formed), thanks to nature's catalytic molecules, the enzymes.

Enzymes are a subclass of biopolymers called proteins. They consist of very long molecules called polypeptides, which are formed by stringing amino acids together. In general they are huge molecules, with molecular masses in the range between

14000 and 8 * 10^6 μ. Needless to say, the structures and stereochemistries of these macromolecules are very complex. The enzyme, chymotrypsin, which is small as enzymes go, is shown in Figure. In the figure, gray balls are carbon, red are oxygen, blue are nitrogen, and yellow are sulfur atoms; hydrogen atoms are not shown. It is estimated that the human body contains more than 2000 enzymes, each with its own unique structure, each catalyzing only a single reaction or closely related set of reactions. Enzymes are extremely selective—specific—in their catalysis.

We will have more to say about enzyme structure in a subsequent chapter. In this section, our focus will be on the kinetics of enzyme catalysis. Despite the vast number of enzyme structures and the equally vast number of reactions catalyzed, most enzyme-catalyzed processes exhibit essentially the same kinetics. Considering the complexity of the molecules and reactions, the kinetics is relatively simple. We approach it first from the experimental standpoint, which produces an observed rate law; we then address the question of mechanism. Equation shows a typical enzyme catalyzed reaction in generic form.

$$\text{Substrate} \xrightarrow{E} \text{Products}$$

Here, E stands for the enzyme, and is written over the arrow to indicate that the reaction is catalyzed. The reactant molecule on which the enzyme works is called the *substrate.* Although there may be more than one reactant molecule, in most cases only one of them interacts directly with the enzyme. We will condense 15–9–1 to the form in 15–9–2.

$$S \xrightarrow{E} P$$

Kinetics studies of enzyme catalyzed reactions usually involve the measurement of initial rates, for reasons that need not concern us here. Throughout the discussion that follows, it will be assumed that the word, rate, stands for initial rate. The general rate law, applicable to most enzyme catalyzed processes for which studies have been done, is in 15–9–3.

$$\text{Rate} = k_{obs}[E]_T$$

Here $[E]_T$ stands for the total concentration of enzyme

present in the solution, and k_{obs} is the observed first-order rate constant. The observed rate constant exhibits an interesting dependence on the concentration of substrate, S. When [S] is small, k_{obs} increases nearly linearly with [S]; in other words, the rate is first order in S. However, as larger and larger concentrations of S are used, the increases in k_{obs} become smaller and smaller; k_{obs} eventually levels off to a constant value that is independent of further increases in [S].

At this point, the rate is zero-order in S. The relationship between k_{obs} and [S] is shown in Figure. To an experienced kineticist, this behaviour is known as *saturation kinetics*, and can be explained by a rate law of the form in 15–9–4.

$$\text{Rate} = k_{obs}[E]_T = A[E]_T[S]/(B + [S])$$

Here A and B are constants that are presumably related to rate constants of the elementary steps in the reaction mechanism. We interpret the observed kinetics in terms of 15-9-4 as follows. When [S] is small, B is the dominant term in the denominator. Under these conditions, 15–9–4 simplifies to

$$\text{Rate} = A[E]_T[S]/B$$

Thus the rate is first order in E, first order in S, with k_{obs} = A/B. When [S] is large, it is the dominant term in the denominator, and 15–9–4 simplifies to

$$\text{Rate} = A[E]_T$$

The rate is first order in E, zero-order in S, with k_{obs} = A. The form of the expression in is consistent with the observed kinetics. For historical reasons, enzyme-substrate systems that obey the kinetic form in 15–9–4 are said to exhibit *Michaelis-Menten kinetics*, after two early investigators in the field. The constant, B, is called the Michaelis constant, and is usually given the symbol, K_M. We will not use this symbol, however, because it implies that B is an equilibrium constant. Instead, we will use the symbol M. We make the switch at this point:

$$\text{Rate} = A[E]_T[S]/(M + [S])$$

$$k_{obs} = A[S]/(M + [S])$$

The plot of Rate versus substrate concentration nicely shows the saturation behaviour of the kinetics; unfortunately, it is not possible to obtain accurate values for A and M from this plot. However, there are two rearrangements of 15–9–8 that allow the experimental data to be plotted in linear form. The first of these is obtained by reciprocating both sides of 15–9–8:

$$1/k_{obs} = M/A[S] + 1/A$$

A plot of $1/k_{obs}$ versus 1/[S] should be linear with slope M/ A and intercept 1/A. Figure shows such a plot for the data of Figure. A reciprocal plot of this type is called a Lineweaver-Burk plot. The values of A and M obtained from are $A = 0.01\ s^{-1}$, M = 0.0055 M. The second linear form is obtained from 15–9–8 by, first, multiplication of both sides by k_{obs};

$$(15\text{–}9\text{–}10)\text{: } 1 = k_{obs}M/A[S] + k_{obs}/A$$

second, multiplication by A;

$$(15\text{–}9\text{–}11)\text{: } A = k_{obs}M/[S] + k_{obs}$$

and third, subtraction of the first term on the right from both sides.

$$(15\text{–}9\text{–}12)\text{: } k_{obs} = A - k_{obs}M/[S]$$

A plot of k_{obs} versus $k_{obs}/[S]$ should be linear with slope - M and intercept A. This plot, called an Eadie-Hofstee Plot, is shown in. It produces M = 0.0055 M, $A = 0.01\ s^{-1}$.

The physical interpretation of Michaelis-Menten (saturation) kinetics goes something as follows. First, the rate is first order in both enzyme and substrate when [S] is small. This implies that 1 molecule of enzyme and 1 molecule of substrate must collide and interact in order to form products. Second, the rate is first order in E and zero-order in S when S is large.

The implication of this observation is that the substrate forms an adduct of some type with the enzyme, which we might symbolize as ES. This adduct then reacts further to form products. When [S] becomes large enough, Le Chatelier's Principle requires that all of the free enzyme is converted to the adduct. Once this happens, the rate no longer increases with increases in [S], because [ES] has reached its maximum possible value. This type of reasoning led biochemists to the simplest mechanism consistent with the observations:

$$E + S = ES\ [k_1, k_{-1}]$$

$$ES \rightarrow P + E\ [k_{cat}]$$

We must now develop the rate law implied by this mechanism, to see that it matches what is observed. It is not legitimate to make assumptions at the outset about the relative magnitudes of the rate constants, because we have no information about that. The development of the rate law must be carried out as generally as possible, then examined for limiting behaviour. We use the so–called *steady state approximation*. The essence of this approximation is that *the concentration of the intermediate adduct, ES, stays essentially constant as reaction proceeds.* This idea is fairly simple to implement. First, we recognize that the rate of formation of products, P, is given by,

$$\text{Rate} = d[P]/dt = k_{cat}[ES]$$

This rate law is strictly correct and always true within the context of the mechanism in 15–9–13. However, it is difficult to use because [ES] is not accessible. As usual, we need to express the rate law in terms of measurable concentrations: those of E and S. Further work is required. Now we invoke the steady state assumption, which is that,

$$d[ES]/dt = 0$$

This is a mathematical way to say what we said above: the concentration of ES remains constant as reaction proceeds. We now use the elementary steps to obtain an expression for [ES]:

$$d[ES]/dt = 0 = k_1[E][S]–k_{-1}[ES]–k_{cat}[ES]$$

Solving for [ES] gives,

$$[ES] = k_1[E][S]/(k_{cat}+k_{-1})$$

which can be substituted for [ES] in the rate law, 15-9-14.

$$\text{Rate} = k_1k_{cat}[E][S]/(k_{cat} + k_{-1})$$

We are close, but not quite done. Recall that the experimental rate law is written in terms of the total enzyme concentration, $[E]_T$. Interpreted in terms of our mechanism,

$$[E]_T = [E] + [ES]$$

In words, the enzyme is partitioned between the unbound and substrate bound forms. We cannot at the outset expect to know how much enzyme is in each form. However, we do know the total amount; it is the amount we put into the solution. We must therefore obtain an expression for [E] in terms of $[E]_T$, which can be done using 15-9-17 and 15–9–19:

$$[E]_T = [E] + [ES] = [E] + k_1[E][S]/(k_{cat} + k_{-1})$$

Solving for [E] gives,

$$[E] = [E]_T/\{1 + (k_1[S]/(k_{cat} + k_{-1})\}$$

Substituting this expression into the rate law, 15-9-18, and doing some simplifying arithmetic gives

$$Rate = k_{cat}[E]_T[S]/\{[S] + (k_{cat} + k_{-1})/k_1\}$$

This has a familiar look. In fact, if we identify k_{cat} with A, and $(k_{cat} + k_{-1})/k_1$ with M, we obtain

$$Rate = A[E]_T[S]/\{M + [S]\}$$

which is identical in form with the experimental law.

Within the context of mechanism, we can draw the following conclusions:

- The value of A gives a direct measure of the rate constant k_{cat}, which governs the rate at which the enzyme substrate adduct reacts to form products.
- The Michaelis constant, M, is seen to be $(k_{cat} + k_{-1})/k_1$. This gives the ratio of the rate of consumption of ES to its rate of production. Importantly, it is clear that M is not an equilibrium constant, because it involves the sum of rate constants from two different elementary processes.

MECHANISMS OF SUBSTITUTION REACTIONS

A substitution reaction is represented generally in equation 1:

$$(15–10–1): A\text{-}X + Y \rightarrow A\text{-}Y + X$$

This reaction type, in which a group Y replaces a group X in a molecule, is widespread in chemistry. Several examples of substitution reactions are presented in equations 2-5.

$$(15–10–2): H–CN + F^- \rightarrow HF + CN^-$$

$$(15–10–3): CH_3CH_2Cl + OH^- \rightarrow CH_3CH_2OH + Cl^-$$

$$(15–10–4): (NH_3)_5CoCl^{2+} + H_2O \rightarrow (NH_3)_5CoOH_2^{3+} + Cl^-$$

$$(15–10–5): PCl_5 + H_2O \rightarrow P(OH)_5 + HCl$$

Reaction (2) is a Bronsted-Lowry acid base reaction, discussed. It is also a substitution reaction in which one base,

F^-, replaces another base, CN^-, in forming a bond to the proton, H+. Reaction (15–10–3) shows the formation of an alcohol from an alkyl halide by substitution of a hydroxide anion for a chloride anion at carbon. Reaction (15–10–4) involves the replacement of one *ligand* by another in a transition metal complex. We will discuss these in more detail below. Finally, (15–10–5) is a double substitution (or double displacement) process in which OH^- and Cl^- substitute for each other. Because substitution reactions are so common, they have attracted the attention of chemists for many years. Of particular interest has been the mechanism(s) by which they take place. What are the details of the process the net result of which is replacement of X by Y? Do all substitution reactions occur in the same way, or are there different pathways that depend in some understandable way on the structures of A, X, and Y?

In this section we will explore these questions and attempt to supply some answers. As we will see, kinetics studies have played the central role in the discovery of mechanism. Our discussion will parallel the historical development of the subject, beginning with the mechanisms of substitution reactions at carbon.

SUBSTITUTION REACTIONS AT CARBON

The replacement of a halide by a hydroxide anion is shown in 15-10-6:

$$(15–10–6): R–Cl + OH^- \rightarrow ROH + Cl^-$$

R represents a carbon-containing group of atoms that may be simple (for example, CH_3) or more complex (for example, $(CH_3)_3C^-$). (15–10–3) is a specific example of the general process in (15–10–6). It has been shown via many research studies over many years that (15–10–6) occurs by two different pathways, depending on the nature of the carbon-containing group R.

The Second-Order Pathway

Consider first the kinetics of (15-10-3), in which R can be identified with the CH_3CH_2– portion of the reactant, CH_3CH_2Cl. The reaction can be studied under pseudo-order conditions in

CH_3CH_2Cl. Recall that this means that the concentration of OH- is maintained at a value at least 10 times larger than the concentration of CH_3CH_2Cl, so that $[OH^-]$ changes imperceptibly during reaction. Studied in this manner, the kinetics are first-order in CH_3CH_2Cl, with a pseudo-first order rate constant, k′, that is directly proportional to the concentration of OH^-. The experimental rate law thus has the form in (15-10-7).

$$\text{Rate} = k_{obs}[CH_3CH_2Cl] \text{ where } k_{obs} = k[OH^-]$$

Thus (15–10–3) is experimentally observed to follow second-order kinetics. From this experimental basis, the chemist must then speculate as to the possible mechanisms that could give rise to a rate law of this form.

In other words, the chemist must propose a theory about the elementary steps involved in the reaction process. Reaction seems quite simple, so it is conceivable that it could occur in a single step in which the Cl^- ion leaves simultaneously with attack by the OH^- ion. In other words, (15–10–3) could be an elementary process. As we know, such processes have rate laws that follow directly from their stoichiometries. Thus, IF were to occur in the single step (15-10-7), it would necessarily obey the rate law.

$$CH_3CH_2Cl + OH^- \rightarrow HO{-}CH_3CH_2{-}Cl \rightarrow CH_3CH_2OH +$$
$$Cl^- [k_1] \rightarrow \text{Rate} = -d[RCl]/dt = k_1 [RCl] [OH^-]$$

This agrees with the experimental rate law, and so the single-step mechanism is consistent with experiment. However, the thorough chemist would not stop here. Rather, s/he would ask whether there are other mechanisms that are consistent with the experimentally observed second-order kinetics. In fact, there is at least one other mechanism that should be considered. It is the two-step mechanism:

$$CH_3CH_2Cl + OH^- \Leftrightarrow CH_3CH_2(OH)Cl^- [k_1, k_{-1}]$$

$$CH_3CH_2(OH)Cl^- \rightarrow CH_3CH_2OH + Cl^- [k_2]$$

In the first step of this process, the OH– group forms a bond to the carbon atom that is attached to Cl^-, forming an intermediate in which this carbon atom is attached to 5 other groups.

In the second step, the bond to the Cl^- ion breaks, generating the products. The reverse arrow in the first step is included to recognize that the just-formed bond to the OH– ion could also break, regenerating the reactants. A steady–state treatment of the mechanism, in which the concentration of the intermediate is assumed to stay constant during the course of reaction, yields the theoretical rate law in (15–10–10):

$$Rate = k_1k_2[CH_3CH_2Cl][OH-]/(k_{-1} + k_2)$$

The 2-step mechanism in (15–10–9) also predicts second-order kinetics. In this case, the experimental k_{obs} would be identified with the expression $k_1k_2/(k_{-1}+k_2)$, which involves the rate constants of all of the elementary steps in mechanism. The important point here is that the experimental rate law does not enable us to choose between the one-step elementary mechanism and the more elaborate two-step mechanism.

Indeed, there are probably other even more elaborate mechanisms that also predict second-order kinetics. Any Mechanism That Predicts A Rate Law Having The Same Form As The Experimental Rate Law Must Be Considered As Possible. Experiments of other types must be designed in an attempt to narrow the field of possible mechanisms. Many such experiments have been designed and carried out. Based on the results, most chemists now believe that reaction (15–10–3) occurs by the elementary mechanism.

Because reaction is only a specific example of reactions of type, is it then reasonable to conclude that all reactions of type (15–10–6) occur by the same single-step process? As is so often the case in science, the situation is not nearly so clean and simple! Thus, although many substitution reactions at carbon show second-order kinetics and are therefore consistent with the one-step mechanism, other reactions show rate laws that are first order in RCl and zero-order in OH–. An example.

$$(CH_3)_3CCl + OH^- \rightarrow (CH_3)_3COH + Cl^-$$

Kinetics studies of (15-10-11) yield the experimental rate law

$$Rate = k_{obs}[(CH_3)_3CCl]$$

Interestingly, the rate of (15–10–11) is independent of how

much OH^- is made availa and that the slow step does not involve the incoming group OH^-proposed to explain this behaviour? First-order kinetics implies that the process occurs in at least two steps, and that the slow step does not involve the incoming group OH^-. One mechanism that produces a rate law of the form of (15-10-12) is given in (15-10-13).

$$(CH_3)_3CCl \rightarrow (CH_3)_3C^+ + Cl^- \ [k_1 \text{ (slow)}]$$

$$(CH_3)_3C^+ + OH^- \rightarrow (CH_3)_3COH \ [k_2]$$

Because the overall rate is determined by the rate of the slowest (rate-determining) step, the theoretical rate law for mechanism (15–10–13) is

$$\text{Rate} = k_1[(CH_3)_3CCl]$$

This is consistent with the experimental rate law.

Mechanism is oversimplified, however, because it ignores the possibilty that the Cl– ion could reattach to the carbon species, $(CH_3)_3C^+$. We can fix this by including the reverse process. The amended mechanism is in (15-10-15).

$$(CH_3)_3CCl \Leftrightarrow (CH_3)_3C^+ + Cl^- \ [k_1, k_{-1}]$$

$$(CH_3)_3C^+ + OH^- \rightarrow (CH_3)_3COH \ [k_2]$$

A steady-state treatment on the intermediate, $(CH_3)_3C^+$, gives the rate law in (15–10–16).

$$\text{Rate} = k_1k_2[(CH_3)_3Cl][OH^-]/(k_{-1}[Cl^-] + k_2[OH^-])$$

Allowing for the reverse of the first step therefore greatly complicates the theoretical rate law. The most noticeable feature of (15–10–16) is that it is no longer obvious that the kinetics are simple first-order in $(CH_3)_3CCl$. Instead, the overall reaction order depends on the relative size of the 2 terms in the denominator. Under conditions where $k_2[OH^-] \gg k_{-1}[Cl^-]$, the $k_2[OH^-]$ term dominates and will cancel with the $k_2[OH^-]$ term in the numerator. The resulting rate law is in:

$$\text{Rate} = k_1[(CH_3)_3Cl]$$

This is consistent with the observed first order behaviour. However, under conditons where $k_2[OH^-] \ll k_{-1}[Cl^-]$, the latter term will dominate the denominator, and the rate law becomes

$$\text{Rate} = k_1k_2[(CH_3)_3CCl][OH^-]/k_{-1}[Cl^-]$$

Under conditions of small [OH–], the rate should be first order in $(CH_3)_3Cl$, first order in OH^-, and inverse first order in Cl^-. Indeed, this suggests that this mechanism could be tested by studying the kinetics using a small $[OH^-]$. Observation of an inverse dependence on $[Cl^-]$ would provide support for mechanism.

In summary, to account for the varied kinetics behaviour displayed by reactions of type (15–10–6), chemists have proposed two limiting mechanisms. These are shown below with their traditional names:

The S_N2 Mechanism:

$$R–X + Y \rightarrow (Y—R—X) \rightarrow R–Y + X$$

The S_N1 Mechanism:

$$R–X \rightarrow R + X$$

$$R + Y \rightarrow R–Y$$

The abbreviation, S_N2, stands for *substitution nucleophilic second-order*. This name is a mouthful, but it is easily explained. "Substitution" and "second-order" are self-explanatory.

The adjective nucleophilic (nucleus-loving) is used to describe an atom, group of atoms, or ion that is attracted to a positive or partially-positive center in a molecule. Both X and Y are nucleophilic groups that compete for the positive center, R+.

In the S_N2 mechanism, Y attacks the carbon atom in R–X and forms a weak bond. Simultaneously, the bond between R and X weakens. The result is a transition state, Y—R—X, in which both the leaving group (X) and the entering group (Y) are weakly bonded to R. The reaction is completed when the bond to X is severed and replaced with a full bond to Y.

The reaction thus occurs in a single elementary step and therefore must display second-order kinetics, consistent with its stoichiometry. The abbreviation, S_N1, stands for *substitution nucleophilic first-order*. The name suggests a substitution involving the competition of 2 nucleophiles for a positive center that displays first order kinetics.

SUBSTITUTION IN TRANSITION METAL COMPLEXES

Elements in the d block of the periodic table, sandwiched between the s block elements at the left and the p-block elements at the right, are called transition metals. These elements form a class of compounds in which the transition metal is bonded to a number of ions or molecules via covalent bonds. When attached to a transition metal ion, such ions or molecules are called ligands. Most ligands are also Bronsted-Lowry bases, so that we can recognize them by the availability of at least one lone pair of electrons. In fact, it is a lone pair of electrons on the ligand that is used to form the bond to the transition metal ion. Some examples of transition metal complexes are given below.

$Fe(CN)_5(NO)^{3-}$. The transition metal ion is Fe^{2+}. It is attached to 6 ligands, five of them cyanide anions (CN^-) and the sixth a nitric oxide molecule. The complex has a net charge of 3 , obtained by summing the charges on the metal ion and the ligands. The six ligands are arranged about the metal ion in an octahedron. The complex is brown in colour.

$Co(NH_3)_5Cl^{2+}$. The transition metal ion is Co^{3+}. It is attached to 6 ligands, five of them ammonia molecules (NH_3) and the sixth a chloride anion. The net charge of 2+ comes from summing the metal charge (3+) and the chloride charge (1-). The six ligands are arranged about the metal ion in an octahedron. The complex is green.

$CoCl_4^{2-}$. The metal ion is Co^{2+}, which is attached to four Cl^- ligands arranged about it in a tetrahedron. The overall charge is the sum of the ion charge (2+) and the chloride charges (4–). The complex is intensely blue in colour.

$Ni(CN)_4^{2-}$. The metal ion is Ni^{2+}, attached to four cyanide ligands which form a square planar arrangement. The complex is yellow in colour.

These few examples illustrate some generalities about transition metal complexes:

- Many of them are coloured.
- Common numbers of ligands are 6 and 4.
- Six ligands almost invariably form an octahedron. Four ligands form either a tetrahedron or a square plane.

The great variety of colors exhibited by transition metal complexes has captured the interest of chemists for centuries. Long ago it was observed that colour changes sometimes occur when a complex is put in the presence of another potential ligand different from those already contained in it. The colour changes are the result of substitution of one ligand by another, as for example in equation (15-10-19):

$$CoCl_4^{2-}(\text{blue}) + 6H_2O \rightarrow Co(H_2O)_6^{2+}(\text{pink}) + 4\ Cl^-$$

The mechanisms of ligand substitution reactions have been intensely investigated since the 1950s. The rate at which one ligand substitutes for another depends tremendously on what the metal is, and on what the other ligands are. In complexes of Cu^{2+}, one ligand can substitute for another at a rate of up to 10^{10} times per second, whereas in complexes of Co^{3+} and Cr^{3+}, substitution can take hours or days. It is not surprising that study of complexes of the latter two ions has been very popular, and quite a bit is now known about their mechanisms of substitution.

Substitution at Co^{3+}. Consider reaction (15–10–20), in which a water ligand is replaced by a chloride ion in aqueous solution:

$$Co(NH_3)_5(H_2O)^{3+} + Cl^- \rightarrow Co(NH_3)_5Cl^{2+} + H_2O$$

Kinetics studies of this reaction give the rate law, Rate = $k_{obs}[Co(NH_3)_5(H_2O)][Cl^-]$. This is consistent with several mechanisms, shown below:

Mechanism 1, one-step:

$$Co(NH_3)_5(H_2O)^{3+} + Cl^- \rightarrow Co(NH_3)_5(H_2O)(Cl)^{2+} \rightarrow Co(NH_3)_5Cl^{2+} + H_2O$$

This mechanism implies the rate law, Rate = $k_1[Co(NH_3)_5(H_2O)^{3+}][Cl^-]$, where k_1 is the rate constant for the single elementary step process shown.

Mechanism 2, two-step:

$$Co(NH_3)_5(H_2O)^{3+} \Leftrightarrow Co(NH_3)_5^{3+} + H_2O\ [k_1\ (\text{slow}),\ k_{-1}\ (\text{fast})]$$

$$Co(NH_3)_5^{3+} + Cl- \rightarrow Co(NH_3)_5Cl^{2+}\ [k_2,\ \text{fast}]$$

The steady-state assumption for the intermediate, $Co(NH_3)_5^{3+}$, yields the rate law in (15-10-23).

$$Rate = k_1k_2[Co(NH_3)_5(H_2O)^{3+}][Cl-]/(k_{-1}[H_2O] + k_2[Cl-])$$

In aqueous solution, $[H_2O]$ is huge and constant, so the first denominator term is expected to be much larger than the second term. The rate law then simplifies to

$$Rate = k_1k_2[Co][Cl-]/k_{-1}[H_2O] = k_{theor}[Co][Cl-]$$

which agrees with the observed rate law if $k_{obs} = k_1k_2/k_{-1}[H_2O]$. This is called *the Dissociative Mechanism,* because the outgoing ligand first dissociates from the metal before the incoming ligand bonds.

Mechanism 3, two-step:

$$Co(NH_3)_5(H_2O)^{3+} + Cl^- \Leftrightarrow Co(NH_3)_5(H_2O)(Cl)^{2+} \; [k_1, k_1]$$

$$Co(NH_3)_5(H_2O)(Cl)^{2+} \rightarrow Co(NH_3)_5Cl^{2+} + H_2O \; [k_2]$$

The implied rate law is,

$$Rate = k_1k_2[Co(NH_3)_5(H_2O)^{3+}][Cl^-]/(k_{-1} + k_2)$$

which agrees with the observed rate law if $k_{obs} = k_1k_2/(k_{-1}+k_2)$. This is called the *Associative Mechanism,* because the incoming ligand is proposed to first associate with the metal before the outgoing ligand leaves.

Mechanism 4, two-step:

$$Co(NH_3)_5(H_2O)^{3+} + Cl^- \Leftrightarrow [Co(NH_3)_5(H_2O)\cdots Cl]^{2+} \; [K_{eq}]$$

$$[Co(NH_3)_5(H_2O)\cdots Cl]^{2+} \rightarrow Co(NH_3)_5Cl^{2+} + H_2O \; [k_2, \text{slow}]$$

The implied rate law for this mechanism is,

$$Rate = k_2K_{eq}[Co]_T[Cl^-]/(1 + K_{eq}[Cl^-])$$

where $[Co]_T$ represents the sum of the concentrations of $Co(NH_3)_5(H_2O)^{3+}$ and $[Co(NH_3)_5(H_2O]\cdots Cl]^{2+}$. This, too, predicts second order kinetics under conditions where $K_{eq}[Cl^-] \ll 1$. This mechanism is called the *Interchange mechanism,* because the RDS involves an interchange of the incoming and outgoing ligands at the metal.

How are these mechanisms to be distinguished? Examination of the implied rate laws provides the answer. Both the dissociative and the interchange mechanisms predict that at sufficiently high concentration of the incoming ligand, the rate should become zero-order in incoming ligand. This is because the denominator term involving the incoming ligand should eventually become large enough to dominate, and would then cancel the incoming ligand factor in the numerator.

This behaviour has been experimentally observed in the reaction of a few complexes of Co^{3+} with anions, and on this basis the one-step and associative mechanisms have been tentatively ruled out for ligand substitution at this metal ion. It is now accepted by most chemists that substitution in Co^{3+} complexes occurs by the interchange mechanism because, despite much work, an intermediate of the type proposed in the dissociative mechanism has not been experimentally observed. Substitution at Fe^{2+}. In some ligand environments, substitution reactions at the Fe^{2+} ion are slow enough to be conveniently measured at room temperature. The complex, $Fe(CN)_5X^{3-}$, where X is a variety of neutral molecules, is particularly convenient to study because the CN^- ligands are very difficult to replace. Reactions in which X is replaced by some other group, 0

$$Fe(CN)_5X^{3-} + Y \rightarrow Fe(CN)_5Y^{3-} + X$$

It has been shown experimentally that the reaction obeys a rate law of the form in (15–10–30):

$$\text{Rate} = k_{obs}[Fe] \text{ where } k_{obs} = A[Y]/(B[X] + C[Y])$$

where [Fe] is shorthand for the reactant iron complex. It is thought that substitutions (15–10–29) occur by the dissociative mechanism in (15–10–31).

$$Fe(CN)_5X^{3-} \Leftrightarrow Fe(CN)_5{}^{3-} + X \ [k_1 \text{ (slow)}, k_{-1}\text{(fast)}]$$
$$Fe(CN)_5{}^{3-} + Y \Leftrightarrow Fe(CN)_5Y^{3-} \ [k_2 \text{ (fast)}, k_{-2}\text{(slow)}]$$

A steady-state assumption applied to the intermediate, $Fe(CN)_5{}^{3-}$, produces the rate law (15–10–32), which has the same form as the observed rate law.

$$\text{Rate} = (k_1k_2[Y] + k_{-1}k_{-2}[X])/(k_{-1}[X] + k_2[Y])\{[Fe]-[Fe]_{equil}\}$$

Because the reaction is reversible, the rate law is expressed in terms of the deviation of the concentration of $Fe(CN)_5X^{3-}$ from its equilibrium value. "Rate" then represents the rate of approach of the reaction to equilibrium.

By studying the manner in which k_{obs} varies with [X] and [Y], it is possible to assign values to the rate constants k_1, k_{-1}, k_2, and k_{-2}.

Substitution at Pt^{2+}. Most complexes of Pt^{2+} have four ligands arranged in a square planar arrangement around the Pt^{2+} ion. This leaves the Pt^{2+} center open to attack from above and below the plane of the complex, and suggests that an associative substitution mechanism might be possible.

FUNDAMENTAL SUBSTITUTION MECHANISMS

There are four mechanisms that have been proposed to account for the details of substitution reactions. These are presented below.

Concerted One-Step Substitution,

$$R\text{–}X + Y \rightarrow RY + X \; [k]$$

Rate law: Rate = k[RX][Y] (Second-order kinetics)

Associative Substitution,

$$R\text{–}X + Y \Leftrightarrow Y\text{–}R\text{–}X \; [k_1, k_{-1}]$$

$$Y\text{–}R\text{–}X \rightarrow R\text{–}Y + X \; [k_2]$$

$$\text{Rate} = k_1k_2[RX][Y]/(k_{-1} + k_2) \text{ (Second-order kinetics)}$$

Dissociative Substitution,

$$R\text{-}X \Leftrightarrow R + X \; [k_1, k_{-1}]$$

$$R + Y \rightarrow RY \; [k_2]$$

$$\text{Rate} = k_1k_2[R\text{–}X][Y]/(k_{-1}[X] + k_2[Y])$$

Interchange,

$$R\text{-}X + Y \Leftrightarrow R\text{-}X...Y \; [K_1]$$

$$R\text{–}X...Y \rightarrow RY + X \; [k_2]$$

$$\text{Rate} = k_2K_1[R\text{–}X]_T[Y]/(1 + K_1[Y])$$

Supplement: Picosecond Reactions; Laser Promoted Processes.

Supplement: Kinetics, Mechanisms, and Occam's Razor.

APPLICATIONS

- Express the rate of the following reaction in terms of

the change in concentration per unit time of each participating species:

$$N_2 + 3H_2 \rightarrow 2NH_3$$

- Complete the following statements for the reaction

$$2N_2O_5(g) \rightarrow 4NO_2(g) + O_2(g).$$

  - The rate of decomposition of N_2O_5 is _____ times the rate of formation of O_2.
  - The rate of formation of NO_2 is _____ times the rate of decomposition of N_2O_5.
  - The rate of formation of NO_2 is _____ times the rate of formation of O_2.
- For the reaction in Application 1, what is the rate of appearance of ammonia if the rate of consumption of nitrogen is 0.01 moles L^{-1} s^{-1}? What is the reaction rate?
- In the reaction 3 $ClO^-(aq) \rightarrow 2\ Cl^-(aq) + ClO_3^-(aq)$, the rate of formation of Cl^- is 3.6 mol/L–min. What is the rate of reaction of ClO^-? What is the unique rate of reaction?
- Dinitrogen pentoxide, N_2O_5, decomposes by a first–order reaction. What is the initial rate of decomposition of N_2O_5 when 2.00 g of N_2O_5 is confined in a 1.00 L container and heated to 65°C? The rate constant for the reaction is $5.2 * 10^{-3}$ s^{-1} at 65°C.
- When the NO concentration is doubled, the rate of the reaction $2NO(g) + O_2(g) \rightarrow 2NO_2(g)$ increases by a factor of 4. When both the O_2 and NO concentrations are doubled, the rate increases by a factor of 8. What are the reactant orders and overall order of the reaction?
- Write the rate law for the consumption of persulfate ions in the reaction $S_2O_8^{2-}(aq) + 3I^-(aq) \rightarrow 2\ SO_4^{2-}(aq) + I_3^-(aq)$ with respect to each reactant and determine the value of k from the following data.

Experiment	Initial Concentration $S_2O_8^{2-}$ mol/L	Initial Concentration I^- mol/L	Initial Rate (mol $S_2O_8^{2-}$)$L^{-1}s^{-1}$
1	0.15	0.21	1.14

2	0.22	0.21	1.70
3	0.22	0.12	0.98

- The following kinetics data were obtained for the reaction, $2ICl(g) + H_2(g) \rightarrow I_2(g) + 2HCl(g)$.

Experiment	Initial Concentration ICl mmol/L	Initial Concentration H_2 mmol/L	Initial Rate mol $L^{-1}s^{-1}$
1	1.5	1.5	$3.7*10^{-7}$
2	3.0	1.5	$7.4*10^{-7}$
3	3.0	4.5	$22*10^{-7}$
4	4.7	2.7	?

  - Write the rate law for the reaction.
  - Determine the value of the rate constant from the data.
  - Use the data to predict the reaction rate for experiment 4.
- The following data were obtained for the reaction A + B + C → products.

Experimen	$[A]_o$	$[B]_o$	$[C]_o$	Initial Rate mmol $L^{-1}s^{-1}$
1	1.25	1.25	1.25	8.7
2	2.5	1.25	1.25	17.4
3	1.25	3.02	1.25	50.8
4	1.25	3.02	3.75	457
5	3.01	1.00	1.15	?

  - Write the rate laq for the reaction.
  - What is the order of the reaction?
  - Determine the value of the rate constant.
  - Use the data to predict the reaction rate for experiment 5.
- Determine the rate constants for the following first–order reactions
  - A → B, given that the concentration of A decreases to one-half its initial value in 1000 s.

  - $A \rightarrow B$, given that the concentration of A decreases from 0.33 mol*L^{-1} to 0.14 mol*L^{-1} in 47 s.
  - $2A \rightarrow B + C$, given that $[A]_o$ = 0.050 mol*L^{-1} and that after 120 s the concentration of B rises to 0.015 mol*L^{-1}.
- Dinitrogen pentoxide, N_2O_5, decomposes by first-order kinetics with a rate constant of 3.7 * 10^{-5} s^{-1} at 298 K.
  - What is the half-life in hours for the decomposition of N_2O_5 at 298 K?
  - If $[N_2O_5]_o$ = 2.33 * 10^{-2} mol*L^{-1}, what will the concentration of N_2O_5 be after 2.0 h?
  - How much time in minutes will elapse before the N_2O_5 concentration decreases from 2.33 * 10^{-2} mol*L^{-1} to 1.76 * 10^{-2} mol*L^{-1}?
- The half-life for the first-order decomposition of A is 200 s. How much time must elapse for the concentration of A to decrease to a) one-half, b) one–sixteenth, c) one-ninth of its initial concentration?
- Sulfuryl chloride, SO_2Cl_2, decomposes by first-order kinetics with k = 2.81 * 10^{-3} min^{-1} at a certain temperature.
  - What is the half-life for the reaction?
  - Determine the time needed for the concentration of a SO_2Cl_2 sample to decrease to 10% of its initial concentration.
  - If a 14.0–g sample of SO_2Cl_2 is sealed in a 2500–L reaction vessel and heated to the specified temperature, what mass will remain after 1.5 hr?
- For the first-order reaction $A \rightarrow 3B + C$, when $[A]_o$ = 0.015 mol/L, the concentration of B increases to 0.020 mole/L in 3.0 min.
  - What is the rate constant for the reaction?
  - How much more time would be needed for the concentration of B to increase to 0.040 mol/L?
- The following data were collected for the reaction $2N_2O_5(g) \rightarrow 4NO_2(g) + O_2(g)$ at 25 °C.

Time, s	$[N_2O_5]$, 10^{-3} mol/L
0	2.15
4000	1.8
8000	1.64
12000	1.43
16000	1.25

- Plot the data appropriately to determine the order of reaction.
- Determine the rate constant from the graph.

• The following data were collected for the reaction $2HI(g) \rightarrow H_2(g) + I_2(g)$ at 580 °C.

Time, s	[HI], 10^{-3} mol/L
0	1000
1000	112
2000	61
3000	41
4000	31

- Plot the data appropriately to determine the reaction order.
- Determine the rate constant from the graph.

• The half-life for the second-order reaction of a substance A is 50.5 s when $[A]_o$ = 0.84 mol/L. Calculate the time needed for the concentration of A to decrease to a) one-sixteenth; b) one-fourth; c) one-fifth of its original value.

• Determine the rate constant for the following second-order reactions
  - $2A \rightarrow B + 2C$, given that the concentration of A decreases from $2.5 * 10^{-3}$ mol/L to $1.25 * 10^{-3}$ mole/L in 100 s.
  - $3A \rightarrow C + 2D$, given that $[A]_o$ = 0.30 mol/L, and that the concentration of C increases to 0.010 mol/L in 200 s.

- Calculate the activation energy for the conversion of cyclopropane to propene from an Arrhenius plot of the following data:

T, K	k, s^{-1}
750	$1.8 * 10^{-4}$
800	$2.7 * 10^{-3}$
850	$3.0 * 10^{-2}$
900	0.26

What is the value of the rate constant at 600 °C?

- Write the overall reaction for the mechanism below, and identify intermediates:

$$ICl + H_2 \rightarrow HI + HCl$$
$$HI + ICl \rightarrow HCl + I_2$$

- Write the overall reaction for the mechanism below and identify intermediates:

$$Cl_2 \rightarrow 2Cl$$
$$Cl + CO \rightarrow COCl$$
$$COCl + Cl_2 \rightarrow COCl_2 + Cl$$

- Develop the rate law implied by the following mechanism:

$$NO + Br_2 \rightarrow NOBr_2 \text{ (slow)}$$
$$NOBr_2 + NO \rightarrow 2\ NOBr \text{ (fast)}$$

- Develop the rate law implied by the following mechanism:

$$Cl_2 \Leftrightarrow 2Cl \text{ (both fast, equilibrium)}$$
$$CHCl_3 + Cl \rightarrow CCl_3 + HCl \text{ (slow)}$$
$$CCl_3 + Cl \rightarrow CCl_4 \text{ (fast)}$$

- Develop the rate law implied by the following mechanism, proposed for the formation of phosgene from CO and Cl_2:

$$Cl_2 \Leftrightarrow 2\ Cl \text{ (both fast, equilibrium)}$$
$$Cl + CO \Leftrightarrow COCl \text{ (both fast, equilibrium)}$$
$$COCl + Cl_2 \rightarrow COCl_2 + Cl \text{ (slow)}$$

- Three mechanisms have been proposed for the reaction

$$NO_2(g) + CO(g) \rightarrow CO_2(g) + NO(g)$$

- *Mechanism 1:*
  - $NO_2 + CO \rightarrow NO + CO_2$
- *Mechanism 2:*
  - $NO_2 + NO_2 \rightarrow NO + NO_3$ (slow)
  - $NO_2 + NO_2 \Leftrightarrow NO + NO_3$ (fast equil)
- *Mechanism 3:*
  - $NO_3 + CO \rightarrow NO_2 + CO_2$ (fast)
  - $NO_3 + CO \rightarrow NO_2 + CO_2$ (slow)

The observed rate law is Rate = $k[NO_2]^2$. Which mechanisms are feasible?

- For each mechanism in application 15–25, draw a reaction coordinate diagram.
- Critically analyse each of the following statements about elementary processes:
  - At equilibrium, the rate constants of the forward and reverse reactions are equal.
  - For a reaction with a very large equilibrium constant, the rate constant of the reverse reaction is much larger than the rate constant of the forward reaction.
  - The equilibrium constant for the reaction equals the ratio of the forward and reverse rates.
  - For an exothermic process, the rates of forward and reverse reactions are affected in the same way by a rise in temperature.
- For the reversible elementary process $2A \Leftrightarrow B + C$, the rate constant for the forward reaction is 265 $L{*}mol^{-1}{*}min^{-1}$; the rate constant for the reverse reaction is 392 $L{*}mol^{-1}{*}min^{-1}$; the activation energy for the forward reaction is 39.7 kJ/mol; and the activation energy for the reverse reaction is 25.4 kJ/mol.
  - Draw the reaction coordinate diagram.
  - What is the equilibrium constant for the reaction?
  - Is the reaction exo- or endothermic?
  - What effect will an increase in temperature have on the rate and equilibrium constants?
- For the reversible elementary process, $A + B \Leftrightarrow C + D$,

the forward rate constant = 36.4 L*mol^{-1}*h^{-1}; the reverse rate constant = 24.3 L*mol^{-1}*h^{-1}; the forward activation energy is 33.8 kJ/mol; and the reverse activation energy is 45.4 kJ/mol.

- Draw a reaction coordinate diagram.
- What is K_{eq}?
- Is the reaction exo- or endothermic?
- What effect will an increase in temperature have on the rate and equilibrium constants?

- Suppose that a catalyst lowers E_a for a particular reaction from 100 to 50 kJ/mol. By what factor does the reaction rate increase at 400 K, all other factors being equal?
- A reaction rate increases by a factor of 10^3 in the presence of a catalyst at 2500°C. The activation energy of the uncatalysed path is 98 kJ/mol. What is the activation energy of the catalysed path, all other factors being equal?
- The reaction of nitric oxide and hydrogen,

$$2NO(g) + H_2(g) \rightarrow N_2O(g) + H_2O(g)$$

is thought to proceed by the mechanism below.

$$1)\ 2NO \Leftrightarrow N_2O_2\ k_1, k_{-1}$$

$$2)\ N_2O_2 + H_2 \rightarrow N_2O + H_2O\ k_2$$

- The rate law for the reaction is Rate = $k[NO]^2$ at high pressure of H_2. At low H_2 pressure, the rate law is Rate = $k[H_2][NO]^2$. Apply the steady state approximation to the mechanism to obtain the full implied rate law, and show that it agrees with the experimental observations under the indicated conditions.
- For the second-order reaction $A \rightarrow B + C$, the concentration of A falls from 0.040 mol/L to 0.0050 mol/L in 12 h. What is the rate constant?
- All radioactive decay processes follow first-order kinetics. The half-life of the radioactive isotope tritium, ^3H or T, is 12.3 years. How much of a 1.0 mg ample of tritium would remain after 5.2 years?
- The rate law for the reaction $2NO(g) + 2H_2(g) \rightarrow N_2(g)$

+ $2H_2O(g)$ is Rate = $k[NO]^2[H_2]$. The following mechanism has been proposed:

$$1)\ 2NO \rightarrow N_2O_2$$

$$2)\ N_2O_2 + H_2 \rightarrow N_2O + H_2O$$

$$3)\ N_2O + H_2 \rightarrow N_2 + H_2O$$

- Which step is likely to be rate determining?
- Sketch a reaction profile for the reaction, which is exothermic. Label the activation energy for each step and the overall reaction enthalpy.

- Use the following data to determine the activation energy for the reaction of ethyl bromide with hydroxide ions in aqueous solution.

Temperature, °C	k, L/mole-s
24	$1.3 * 10^{-3}$
28	$2.0 * 10^{-3}$
32	$3.0 * 10^{-3}$
26	$4.4 * 10^{-3}$
40	$6.4 * 10^{-3}$

- Explain in narrative form (*i.e.*, in words) why an enzyme catalyzed reaction approaches a maximum velocity when substrate concentration is steadily increased.
- The decarboxylation of a b-keto acid catalyzed by a decarboxylation enzyme can be measured by the rate of CO_2 formation. From the initial rates given in the table, determine the Michaelis-Menten constant for the enzyme and the maximum velocity by a graphical method.

[Keto Acid], M	Initial Velocity, Micromoles CO_2/2 min
2.500	0.588
1.000	0.500
0.714	0.417
0.526	0.370
0.250	0.256

- The hydration of CO_2,

$$CO_2 + H_2O \Leftrightarrow HCO_3^- + H^+$$

is catalyzed by the enzyme, carbonic anhydrase. The kinetics of the forward (hydration) and reverse (dehydration) reactions at pH 7.1, 0.5°C, and $2*10^{-3}$ M phosphate buffer were studied using bovine carbonic anhydrase. Some results follow:

Hydration		**Dehydration**	
1/v, $M^{-1}s$	**$[CO_2]$, mM**	**1/v, $M^{-1}s$**	**$[HCO_3^{-1}]$, mM**
$36*10^3$	1.25	$95*10^3$	2
$20*10^3$	2.5	$45*10^3$	5
$12*10^3$	5	$29*10^3$	10
$6*10^3$	20	$24*10^3$	15

Determine the Michaelis constant and the rate constant, k_2, for the decomposition of the enzyme–substrate complex to form product for:
- The hydration reaction
- The dehydration reaction
- From your results, determine the equilibrium constant for the reaction.

- At pH 7 the measured Michaelis constant and maximum velocity for the enzymatic conversion of fumarate to L-malate,

$$\text{Fumarate} + H_2O \rightarrow \text{L-malate}$$

are $4.0*10^{-6}$ M and $1.3*10^3[E]_o\ s^{-1}$, respectively, where $[E]_o$ is the total molarity of the enzyme. The Michaelis constant and maximum velocity for the reverse reaction are $1.0*10^{-5}$ M and $800[E]_o\ s^{-1}$, respectively. What is the equilibrium constant for the hydration reaction?

- The simple Michaelis-Menton mechanism for an enzmye catalyzed reaction is

$$E + S \Leftrightarrow ES,\ k_1,\ k_{-1}$$
$$ES \rightarrow E + P,\ k_2$$

The following data were obtained:

k_1, k_{-1} very fast

$k_2 = 100\ s^{-1}$, $K_M = 1.0*10^{-4}$ M at 280 K

$k_2 = 200\ s^{-1}$, $K_M = 1.5*10^{-4}$ M at 300 K

- For [S] = 0.10 M and $[E]_o = 1.0*10^{-5}$ M, calculate the rate of formation of product at 280 K.
- Calculate the activation energy for k_2.
- What is the equilibrium constant at 280 K for the formation of the enzyme substrate complex ES form E and S?
- What is the sign and magnitude of DH° for the formation of ES from E and S?

• Explain in simple terms some of the reasons why enzymes are so *efficient* and so *specific*.

• A particular enzyme is present at a concentration of $1.2*10^{-8}$ M. The rate of substrate disappearance under a particular set of concentration and temperature conditions is 0.1 M/min. What is the turnover number of the enzyme?

• For the enzyme in the previous problem, an increase of pH by 1 unit slows the rate at which the enzyme converts substrate to product by a substantial factor. What tentative conclusions can be drawn about the chemical environment at the active site of the enzyme?

• It is generally stated that a catalyst lowers the activation energy for a reaction by providing an alternative mechanism. Provided this is true, what general statement(s) can you make about the temperature dependence of the rate of an enzyme-catalyzed reaction versus that for the uncatalyzed reaction?

• Equation 11 of Kinetics Handout 2 shows the rate law for a reaction catalyzed by an enzyme in the presence of a *competitive inhibitor* of the enzyme. The rate law is obtained from a steady state treatment of the preceding mechanism. Carry out the steady state treatment and show that the rate law is correct.

• The basic mechanism for enzyme catalysis is usually

written as if it involves only 2 steps: E + S ⇔ ES; ES → E + P. The mechanism of action of chymotrypsin is discussed in Kinetics Handout 1. How many substeps are involved in the "second step" for chymotrypsin?

- Suppose that the equilibrium constant for binding of a molecule of substrate to a molecule of enzyme is $1.0*10^3$ M^{-1}. What fraction of the enzyme will be in the form of the enzyme–substrate complex when substrate concentration is 0.001 M. When substrate concentration is 0.01 M. By what factor will the substrate conversion rate increase when [S] is increased from 0.001 to 0.01 M?
- The equilibrium constant for binding of a molecule of substrate to a molecule of enzyme is 10^4 M^{-1}. What is the minimum substrate concentration necessary to guarantee that the enzyme is operating at more than 99% of its maximum conversion rate?
- You have obtained the following data for the destruction of thiamine by an enzyme, both with and without $0.20*10^{-4}$ M inhibitor.

[Thiamine], mM	Velocity (M/min), Uninhibited	Velocity (M/min), Inhibited
0.10	$0.746*10^{-6}$	$0.136*10^{-6}$
0.25	$0.992*10^{-6}$	$0.285*10^{-6}$
0.50	$1.12*10^{-6}$	$0.618*10^{-6}$
1.00	$1.26*10^{-6}$	$0.758*10^{-6}$
2.00	$1.36*10^{-6}$	$1.29*10^{-6}$

  - Make suitable plots and use the V_{max} and K_M values to determine the type of inhibition.
  - Based on these results, determine the binding constant for the inhibitor.

SPREADSHEET APPLICATIONS

- Carry out a finite difference analysis of the kinetics of the following sequence of elementary processes:

$$A \rightarrow B\ k_1 = 0.20\ s^{-1}$$
$$B \rightarrow C\ k_2 = 0.15\ s^{-1}$$

Show how the concentrations of A, B, and C vary with time.

- The following concentration time data were obtained for the reaction

$$A \rightarrow C$$

Use finite difference methods to determine the order of the reaction in reactant A and the value of the rate constant.

Time, s	[A], M
0	0.15
40	0.127
80	0.107
120	0.0906
160	0.0766
200	0.0648
240	0.0547
280	0.0463
320	0.0391
360	0.0331
400	0.028
440	0.0236
48	0.020
520	0.0169
560	0.0143
600	0.0121
640	0.0102
680	0.00862
720	0.00729
760	0.00616
800	0.00521
840	0.0044
880	0.00372
920	0.00315
960	0.00266
1000	0.00225

- The following concentration time data were obtained for the reaction

$$A + 2B \rightarrow C$$

Use finite difference methods to determine the order of the reaction in each of the reactants A and B and the value of the rate constant.

Time, s	[A] for $[B]_o$ = 1.5 M	[A] for $[B]_o$ = 3.0 M
0	0.15	0.15
40	0.117	0.0906
80	0.0906	0.0547
120	0.0704	0.0331
160	0.0547	0.0200
200	0.0425	0.0121
240	0.0331	0.00729
280	0.0257	0.0044
320	0.0200	0.00266
360	0.0155	0.00161
400	0.0121	0.000971
440	0.00938	0.000587
480	0.00729	0.000354
520	0.00567	0.000214
560	0.00440	0.000129
600	0.00342	0.0000781
640	0.00266	0.0000472
680	0.002070	0000285
720	0.00161	0.0000172
760	0.00125	
800	0.000971	
840	0.000755	
880	0.000587	
920	0.000456	
960	0.000354	
1000	0.000275	

- At room temperature, A and B react to form two different products according to the following equations:

$$A + B \rightarrow D \text{ with rate constant } k_1$$
$$A + B \rightarrow F \text{ with rate constant } k_2$$

It is possible to monitor the concentration of D at a wavelength of 540 nm, and the concentration of F at a wavelength of 280 nm. The follo367wing data were obtained in separate experiments, both run under pseudo-order conditions in A. What is the order of the reaction in A? What are the values of k_1 and k_2?

Time s	[D], M	[F], M
0	0.00E+00	0.00E+00
3	4.54E–06	1.34E–05
6	8.94E–06	2.55E–05
9	1.32E–05	3.64E–05
12	1.73E–05	4.63E–05
15	2.13E–05	5.53E–05
18	2.51E–05	6.34E–05
21	2.89E–05	7.07E–05
24	3.25E–05	7.73E–05
27	3.60E–05	8.33E–05
30	3.94E–05	8.88E–05
33	4.26E–05	9.37E–05
36	4.58E–05	9.81E–0
39	4.88E–05	1.02E–04
42	5.18E–05	1.06E–04
45	5.47E–05	1.09E–04
48	5.74E–05	1.12E–04
51	6.01E–05	1.15E–04
54	6.27E–05	1.17E–04
57	6.52E–05	1.19E–04
60	6.76E–05	1.21E–04
63	7.00E–05	1.23E–04
66	7.23E–05	1.25E–04

- The reaction $A + B \Leftrightarrow D$ occurs as a reversible one-step process at room temperature. The kinetics of the net forward process were studied under pseudo-order conditions in A and the reaction was found to be first order in A. The observed rate constant, k_{obs}, was found to vary as follows with [B]. Determine the values of k_1 and k_{-1} for the reaction.

[A], M	[B], M	k_{obs}, s^{-1}
$1.2*10^{-4}$	0.124	$8.65*10^{-3}$
$1.2*10^{-4}$	0.273	$2.01*10^{-2}$
$1.2*10^{-4}$	0.314	$2.19*10^{-2}$
$1.8*10^{-4}$	0.116	$8.03*10^{-3}$
$1.9*10^{-4}$	0.437	$3.58*10^{-}$

Index

G

H

I

J

K

L

M

N

O

P

Q

R

S

T

U

V